A-LEVEL

STUDENT GUIDE

WJEC/Eduqas

Geography

Contemporary themes in geography

Nicky King

Hodder Education, an Hachette Company, Carmelite House, 50 Victoria Embankment, London, EC4Y 0DZ

Orders: please contact Hachette UK Distribution, Hely Hutchinson Centre, Milton Road, Didcot, Oxfordshire, OX11 7HH. Telephone: +44 (0)1235 827827. Email education@hachette.co.uk. Lines are open from 9 a.m. to 5 p.m., Monday to Friday. You can also order through our website: www.hoddereducation.co.uk

© Nicky King 2019

ISBN 978-1-5104-4921-3

First printed 2019

Impression number 5 4 3

Year 2024

This guide has been written specifically to support students preparing for the WJEC/Eduqas A-level examinations. The content has been neither approved nor endorsed by WJEC/Eduqas and remains the sole responsibility of the author.

Typeset by Integra Software Services Pvt. Ltd., Pondicherry, India

Printed by CPI Group (UK) Ltd, Croydon, CR0 4YY

Cover photograph: dabldy/Adobe Stock; other photos p.17 Corel; p.27 David Pearson/Alamy Stock Photo

Hachette UK's policy is to use papers that are natural, renewable and recyclable products and made from wood grown in well-managed forests and other controlled sources. The logging and manufacturing processes are expected to conform to the environmental regulations of the country of origin.

MIX
Paper | Supporting
responsible forestry
FSC™ C104740

Contents

Content Guidance

Questions & Answers

■ Getting the most from this book

Exam tips

Advice on key points in the text to help you learn and recall content, avoid pitfalls, and polish your exam technique in order to boost your grade.

Knowledge check

Rapid-fire questions throughout the Content Guidance section to check your understanding.

Knowledge check answers

1 Turn to the back of the book for the Knowledge check answers.

Summaries

■ Each core topic is rounded off by a bullet-list summary for quick-check reference of what you need to know.

Exam-style questions

Sample student answers

Practise the questions, then look at the student answers that follow.

Commentary on sample student answers

Read the comments (preceded by the icon **e**) showing how many marks each answer would be awarded in the exam and exactly where marks are gained or lost.

Questions & Answers

■ Economic growth and challenge: China

Question 3

'China's physical environment provides more opportunities than constraints for economic development.' Discuss.

e Answers might include the following AO2 points:
- Opportunities and constraints vary spatially — water resources are more abundant to the south.
- Some opportunities may create constraints — the combustion of China's abundant coal reserves, which has fuelled economic growth, has led to increases in the cost of environmental amenity and repair, placing a strain on China's economy.

Extract from student answer

...China benefits from large reserves of coal, which have been used to fuel its economic development. However, these are unevenly distributed and there are limited reserves of high-quality coking coal and anthracite, and both these issues act as a constraint. Although the combustion of China's low-grade coal powers 'the workshop of the world', it also increases emissions of CO_2, leading to increased costs of environmental amenity and repair, and threatening the sustainability of China's economic growth...

...Therefore, China's physical environment presents both opportunities and constraints, and the balance between the two depends on the aspect of the environment under consideration, its location and changes over time. Coal has provided the basis for China's rapid industrialisation, but the constraints associated with fossil fuel use, such as damage to the environment and people's health, are becoming more apparent over time. Conversely, some constraints, such the inaccessible nature of the Qinghai–Tibet Plateau, are now recognised as presenting opportunities for tourism. On balance it can be argued that unless China's physical environment is managed more effectively, for example by using the country's solar and hydropower potential in place of its fossil fuel base, the opportunities will be outweighed by constraints, preventing a sustainable future.

e AO1 band 4 (Eduqas), band 2 (WJEC) Although the response above contains only extracts from the student's full answer, the knowledge of China's physical environment shown in the first paragraph lacks specific locational support and examples. It is worth bearing in mind that an annotated sketch map outlining some of the locations of key resources, landscape and concentrations of hazards (typhoons, earthquakes, sandstorms) may save writing time and earn good AO1 and AO2 credit.

AO2 band 4 (Eduqas), low band 3 (WJEC) The student provides accurate application of knowledge and understanding, but only if this is supported by more detailed evidence (see above) can the student achieve the top band. There

Commentary on the questions

Tips on what you need to do to gain full marks, indicated by the icon **e**

■ About this book

This aim of this guide is to help you succeed in your two chosen themes, selected from the following options:

- Ecosystems
- Economic growth and challenge — India/China/development in an African context
- Energy challenges and dilemmas
- Weather and climate

These optional themes make up:

- section B of Eduqas A-level Geography Component 3: Contemporary themes in geography
- section B of WJEC A-level Geography Unit 4: Contemporary themes in geography

The format of the different examination papers for Section B is summarised as follows:

Specification and paper number	Total marks for Section B	Suggested time spent on Section B	Nature of assessment
Eduqas A-level Component 3	90/128	90 minutes of paper lasting 2 hours 15 minutes	Two extended responses/essays — one from each of your two selected themes
WJEC A2 Unit 4	44/64	80 minutes of paper lasting 2 hours	Two extended responses/essays — one from each of your two selected themes

Section A is based on one compulsory theme — Tectonic hazards. Section A is covered in the *WJEC/Eduqas AS/A-level Geography Student Guide 2: Coastal landscapes/Tectonic hazards* and *Student Guide 3: Glaciated landscapes/Tectonic hazards*. Your centre can choose between Coastal landscapes or Glaciated landscapes. The assessment of answers to Section A is similar to that for Section B, but with a slightly different mark allocation and AO weightings.

This guide has two sections:

- **Content Guidance** — this summarises some of the key information that you need to know to be able to answer the examination questions with a high degree of accuracy and depth. In particular, the meaning of key terms is made clear and some attention is paid to providing details of case study material to help meet the spatial context requirement within the specification.
- **Questions & Answers** — this includes some sample questions similar in style to those you might expect in the exam. There are some sample student responses to these questions as well as detailed analysis, which will give further guidance on what exam markers are looking for to award top marks. The best way to use this book is to read through the relevant theme first before practising the questions. Only refer to the answers and examiner comments after you have attempted the questions.

Content Guidance

■ Ecosystems

The value and distribution of ecosystems

Ecosystems as providers of goods and services

Ecosystem goods are products that can be derived directly from an ecosystem. Examples include timber from trees, nutrients derived from plants and animals, fibres and medicines.

Ecosystem services are benefits that people obtain from ecosystems. These services result from the interactions among organisms and their natural environments, and are important to human wellbeing. Ecosystem services can be categorised as:

- provisioning services (direct products of ecosystems, such as food)
- regulating services (benefits from the natural regulation of, for example, CO_2)
- cultural services (non-material benefits obtained from natural systems, such as aesthetic pleasure from looking at scenery)
- supporting services (e.g. soil formation and nutrient cycling)

Knowledge check 1

Identify synoptic links between ecosystem goods/services and other parts of your A-level course.

> **Exam tip**
>
> When answering a question on the value of ecosystems, note that a given ecosystem good or service might be viewed differently by different groups. Seals and whales can be viewed as a provisioning ecosystem good by the Inuit, whilst whale watching is perceived as a cultural service by tourists.

Exam tip

Remember that AO2 marks are earned by making synoptic links with other parts of the A-level specification, such as the carbon cycle.

Human populations depend on the goods and services provided by the world's ecosystems for their wellbeing and survival (Figure 1). If ecosystems are over-exploited for the short-term provisioning of goods, there will be long-term loss of the valuable services they provide for human wellbeing and ecosystem survival. Unsustainable economic use of ecosystem goods and services places humans at risk.

Knowledge check 2

Can you think of an example of where ecosystem goods or services have been over-exploited by human activity?

> **Exam tip**
>
> Consider spatial variations in the benefits derived from ecosystem goods and services. Growing jatropha (a shrubby tree with oily seeds that can be used to make biodiesel) in Kenya provides an ecosystem good for wealthier EU countries, but may lead to ecosystem degradation in Kenya, resulting in inequalities.

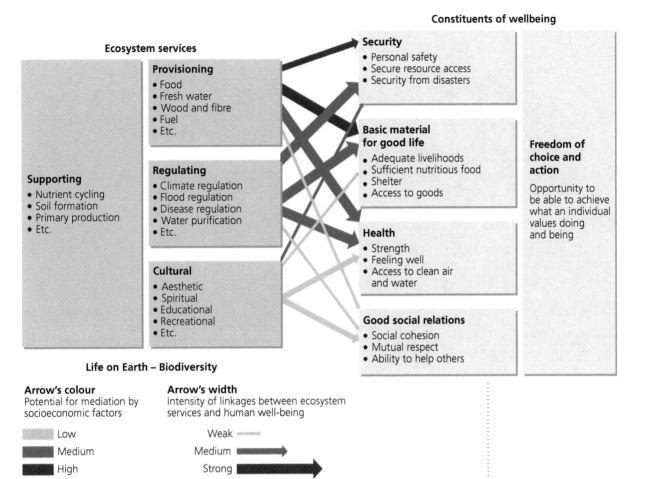

Figure 1 Linkages between biodiversity, ecosystem services and human wellbeing

Distribution of the major global biomes

Figure 2 shows the distribution of **biomes**, which is principally controlled by climate, particularly average temperatures and moisture availability. Different biomes can be grouped according to mean annual temperature and precipitation (Figure 3). Forests, such as the tropical rainforest, located between 10°N and 10°S of the equator in South America, Africa and Asia, are associated with high temperatures and moisture availability. Grasslands (tropical and temperate) are associated with areas of low (often seasonal) moisture availability, but a wider range of temperatures. Areas with extremely low annual precipitation levels correspond with desert biomes.

> A **biome** is a major terrestrial ecosystem of the world.

> **Knowledge check 3**
>
> Will seasonal exchanges of carbon between the atmosphere, biosphere and soil vary more in the tropical rainforest biome or the temperate grassland biome? Why?

> **Exam tip**
>
> Using the correct definitions and appropriate terminology in your answers is important for strengthening AO1 marks (awarded for knowledge and understanding).

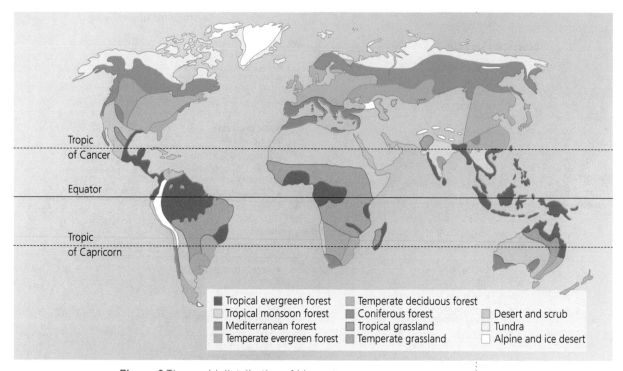

Figure 2 The world distribution of biome types

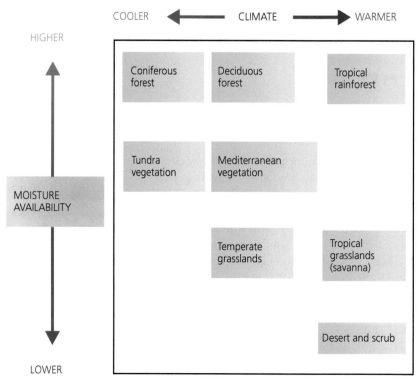

Figure 3 Factors controlling the distribution of biomes

The structure and functioning of ecosystems
The ecosystem concept, including energy flows

An ecosystem is a discrete structural, functional and life-sustaining environmental system. It consists of biotic components (plants, animals and microorganisms) and abiotic components (non-living chemical and physical parts of the environment).

The structure of an ecosystem consists of a series of storage units or **trophic levels**, including autotrophs (primary producers), heterotrophs (consumers) and saprotrophs (decomposers), each of which occupies an ecological niche. Radiant energy from the Sun is the only significant energy source for any ecosystem. Only plants can store energy from the Sun via photosynthesis, so all the other organisms must receive their energy passed along a food chain. Energy is stored within each level and transferred between the levels. The lowest level always consists of primary producers (plants). The second level comprises primary consumers (herbivores), which feed on organisms in the first trophic level. The third level — the secondary consumers (carnivores) — feed on herbivores. The fourth level consists of tertiary consumers, which often include omnivores. The final trophic level is also known as the apex predator. There are never more than five trophic levels. Decomposers feed on all the trophic levels.

Organisms require energy for life functions such as growth, movement and reproduction. Therefore, for all organisms, there must be both a source and an associated loss of energy. The **biomass** decreases at each trophic level because energy is lost at the transfer between stages through respiration and decay. Although the size of the organism generally increases with each trophic level, the number of individuals that can be supported decreases (Figure 4).

A **trophic level** is a feeding level occupied by particular organisms within an ecosystem.

Biomass refers to the total amount of organic matter.

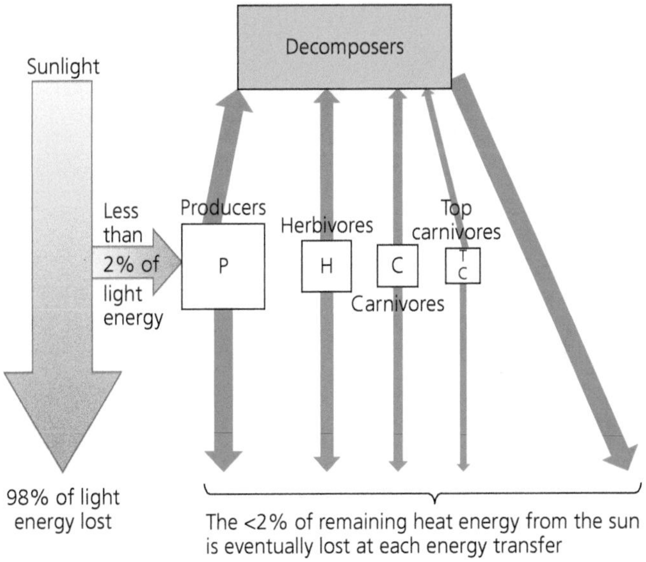

Figure 4 Generalised energy flow and heat loss through an ecosystem

Variations in nutrient cycling between two biomes

Nutrients are the chemical elements and compounds needed for organisms to grow and function. They are stored in soil, litter and biomass. The Gersmehl nutrient cycle model, made up of circles and lines, can be used to summarise nutrient cycling. The size of circles is proportional to the amount of nutrients stored, and the thickness of the lines is proportional to the amount of nutrient flows (Figure 5).

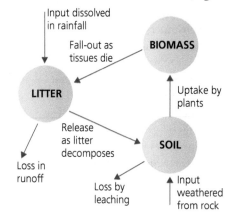

Figure 5 Generalised Gersmehl nutrient cycle model

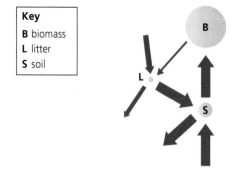

Figure 6 Gersmehl diagram of an equatorial rainforest

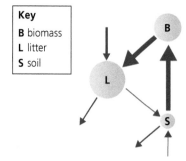

Figure 7 Gersmehl diagram of taiga (northern coniferous forest)

Temperature and precipitation influence the sizes of inputs, stores, flows and outputs. The tropical rainforest biome demonstrates an active cycle of nutrient cycling; the taiga forest biome represents a poor nutrient cycle, with slow replenishment.

The equatorial rainforest biome (Figure 6), with its large above-ground biomass store, contrasts with the smaller biomass store of the taiga forest biome (Figure 7), where growth and biodiversity are lower due to the cold conditions. The litter store for the tropical rainforest biome is small because of rapid decomposition in the hot, wet conditions. This contrasts with the large litter store in the taiga forest biome, where the cold, acidic conditions inhibit the breakdown and decay of litter. The small nutrient soil store in the tropical rainforest biome results from rapid leaching due to high precipitation, whereas the soil store in the taiga forest biome is very small due to the cold conditions, which inhibit **weathering**.

Primary productivity

Levels of **primary productivity** are linked to the presence of limiting factors, including temperature, moisture, light and nutrient availability. Temperature controls the rate of plant metabolism, which, together with light, determines the rate of photosynthesis. Temperatures need to reach a critical level for an ecosystem to function — biological metabolic activity occurs within the range 0–50°C, with an optimum range of 15–25°C. Water is a principal requirement for photosynthesis and is therefore an important influence on levels of primary productivity — in the absence of water the importance of temperature and light diminishes. Temperature and moisture also influence the rate of weathering and therefore nutrient availability from soils. The complexity of, and interconnections between, the limiting factors make it difficult to isolate the importance of one factor alone.

Tables 8 and 9 on pages 47–48 of *WJEC/Eduqas A-level Geography Student Guide 4: Water and carbon cycles/Fieldwork and investigative skills* show the physical factors affecting plant growth in the tropical rainforest and temperate grassland biomes.

> **Exam tip**
>
> Always provide supporting statistics when discussing levels of primary productivity. The mean net primary productivity of tropical rainforest is $2.3\,kg\,m^{-2}\,yr^{-1}$, compared with that of tropical grassland of $0.9\,kg\,m^{-2}\,yr^{-1}$.

Biodiversity under threat

Measures of biodiversity

Ecosystem biodiversity refers to the number and variety of organisms within an area, and has three components: species diversity, ecosystem (or habitat) diversity and genetic diversity. It is often used as a measure of the health of biological systems.

Threats to biodiversity

Some physical and human factors threaten biodiversity both directly and indirectly. Direct threats mostly relate to land management practices, including industry, agriculture, mining, urbanisation and tourism. Indirect threats are mostly associated with human-induced climate change, and include threats from changing temperature and rainfall patterns, rising sea levels, the increased incidence of high-impact storms and ocean acidification.

Weathering is the breakdown (disintegration and decomposition) of rocks *in situ*.

Knowledge check 4

How does nutrient cycling in the temperate grassland biome differ from that in the tropical rainforest and the taiga?

Primary productivity is the rate at which energy can be converted into organic matter. Gross primary productivity is a measure of all the photosynthesis that occurs within an ecosystem. Net primary productivity is the energy fixed in photosynthesis minus the energy lost by respiration.

The biodiversity of some ecosystems is particularly at risk from direct action. Tropical rainforests, such as the Amazon rainforest, are under threat because of deforestation associated with cattle ranching, soy and oil palm cultivation, logging and mineral exploitation. About 95 per cent of all deforestation in the Brazilian Amazon occurs within 50 km of road infrastructure.

The biodiversity of all ecosystems is increasingly at risk from indirect action, particularly the threats associated with **anthropogenic** climate change. These include the increased incidence of high-impact storms (causing habitat destruction), changing temperature and rainfall patterns (which affect the least tolerant species) and sea level rise (which may occur too quickly for species to adjust). Ecosystems at greater distances from human influences are more likely to be threatened by indirect rather than direct action. Due to its relative isolation, low population densities and remoteness, the tundra biome falls into this category; however direct threats associated with mining, settlement and tourism are increasing (p. 18).

Anthropogenic means originating in human activity.

> **Exam tip**
>
> When discussing threats to biodiversity, consider threats at different spatial scales: local (e.g. a pollution incident from one factory), regional (e.g. acid rain or nuclear contamination) and global (e.g. global warming), and over different time scales.

Ecosystems at greatest risk

Tropical rainforests are at risk from climate change, acid rain and invasion by non-native species. They are particularly fragile because of their high biodiversity, and the loss of ecological niches can have dramatic effects on food webs, which are highly specialised. Other concerns include the following:

■ Biomass loss reduces carbon storage and sequestration.
■ Most nutrients are held up in the biomass store (Figure 6), so when deforestation occurs many nutrients are lost.
■ Deforestation reduces inputs of organic matter to the litter and soil stores.
■ The loss of canopy cover results in less interception and more precipitation reaching the forest floor, leading to increased loss of nutrients through leaching and overland flow.

> **Exam tip**
>
> If an exam question asks for an evaluation of ecosystem threats, your analysis could consider spatial scales. Threats can take place at more than one scale and can cross scales. Demand for timber may lead to the regional loss of forest cover, which may increase flood magnitude along a local stretch of river.

Coral reefs are found between latitudes 30° N and 30° S in marine water with a minimum temperature of 18°C, where water is less than 30 m deep, allowing light for photosynthesis. Zooanthellae (algae) survive in these conditions and provide the oxygen required by the corals.

Coral reefs are disappearing because tropical seas have become warmer, causing coral to lose the algae that live within them. In addition, oceans are becoming more acidic as a result of CO_2 emissions associated with human activity. Surface ocean pH has become more acidic by approximately 0.1 units since pre-industrial times. Ocean acidification affects calcium carbonate saturation in ocean waters, thereby making this building block of shells and skeletons less available, which affects the health of corals and other marine organisms.

Depending on their location, many coral reefs, notably the reefs of southeast Asia, are also at risk from direct threats from overfishing and destructive fishing practices (including the use of cyanide and dynamite).

Coral reefs are of particular concern because:
- although they cover less than 0.1 per cent of the ocean floor, they support a quarter of all marine species
- their vertical growth and complexity provide numerous habitats
- they support valuable fish stocks

The bleached coral can recover, but only if cooler water temperatures return and the algae are able to grow again (although there is new research into breeding 'super coral' varieties — more resistant to stresses of warming seas and acidification — in a process known as 'assisted evolution').

Freshwater ecosystems in rivers, lakes and wetlands occupy less than 1 per cent of the Earth's surface, but they have a disproportionately large amount of biodiversity. Over half the world's wetlands have disappeared since 1900. Direct threats from commercial development (including agriculture and tourism facilities), drainage schemes, extraction of minerals and peat, and overfishing have contributed to this loss. The indirect threat of climate change is leading to shallow wetlands being swamped, and some species (e.g. mangrove trees) being submerged and drowned. At the same time, other estuaries, floodplains and marshes are being destroyed through drought.

Exam tip

If an exam question asks you to evaluate which ecosystem is at greatest risk, include criteria such as: the spatial extent of the ecosystem (e.g. wetlands cover only 1 per cent of the Earth's surface) and the ability of an ecosystem to recover (you could integrate the specialised concepts of equilibrium and thresholds when discussing coral reefs).

Conserving biodiversity
Strategies to conserve biodiversity
Strategies to conserve biodiversity range from scientific reserves that prevent access in order to allow total protection to schemes that integrate economic development into conservation (Figure 8).

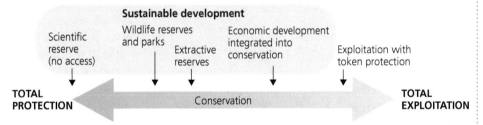

Figure 8 A balancing act is needed between protection and exploitation

Exam tip

Surtsey is an example of a lithosere — one of the local-scale ecosystems you could investigate when studying ecosystem succession.

Surtsey is a volcanic island approximately 32 km from the south coast of Iceland, formed by volcanic eruptions that took place from 1963 to 1967. It has been managed as an IUCN category 1a Strict Nature Reserve since its inception. The fact that the ongoing ecological processes have remained unaffected by human activities makes Surtsey a unique natural laboratory of global significance.

An example of a conservation strategy that involves economic development as a conservation tool is the extractive reserves idea in Brazil, which aims to simultaneously conserve forests and extract their resources in an economically sustainable way. Extractive reserves are not limited to extractive activities, such as nut harvesting and rubber tapping, but can also be used for agricultural activities. Since their inception, more than 3.4 million hectares of Brazilian land are now part of extractive reserves, and several more reserves are in the planning process. In the Alto Juruá reserve, although deforestation is occurring, its frequency is much lower than in neighbouring, non-reserve lands. Where extractive activities involve land clearing, community management has kept the clearing to a minimum.

Conservation issues

Issues associated with conservation include which areas and which species to conserve, the design of areas, *in situ* or *ex situ* approaches, and the legislation and financing of programmes.

Types of conservation area include those with a comprehensive representation of a particular ecosystem type (eco-region approach) and those where the greatest number of species are under imminent threat (hotspots). Eco-regions are smaller than hotspots, which makes them easier to protect.

In situ conservation involves maintaining organisms in their wild state in existing locations, or creating gene pools and banks. *Ex situ* conservation involves moving species and managing them in captivity, which is costly and limited. Both approaches have been used in the case of the Iberian lynx in the dehesa (Spain) and montado (Portugal) ecosystems. Intensive conservation practices were implemented in the wild, including restoration of populations of the lynx's main prey — the rabbit — together with regeneration of rabbit and lynx habitats. This, combined with a captive-breeding programme, led to IUCN revising the status of the Iberian lynx from Critically Endangered to Endangered in 2015.

Conservation is expensive and cannot be left to individual countries. In general, HICs (high income countries, see p. 71) have a higher proportion of protected areas than LICs (low income countries), and have higher conservation costs per km^2 because land prices are higher.

Ecosystems at a local scale
Succession of one ecosystem

Ecosystems are dynamic and can change considerably over time and space. The replacement of some species by others through time is called an ecological succession. Each period or location when a particular type of vegetation is the most important is known as a seral stage. From a 'sterile' area, an ecosystem at the local scale can develop through distinct seres to a **climatic climax community**. The initial physical conditions (water, rock or sand) are important in influencing the specific nature of succession. A hydrosere will develop over time from fresh water, a halosere over time from salt water, a lithosere over time from bare rock and a psammosere over time and space from bare sand (Figure 9).

Psammoseres develop beyond the high-tide mark on a beach. With increasing distance inland, the dunes are older, and the flora and fauna have had longer to colonise and adapt to the environmental conditions. Each stage (or sere) is referred to in terms of its location along the succession, and is associated with certain physical

A **climatic climax community** comprises vegetation that is representative of the final, stable, most complex, self-perpetuating community, which is in equilibrium with the local or regional climate.

A **prisere** is the development of a succession from the initial bare surface to climatic climax community.

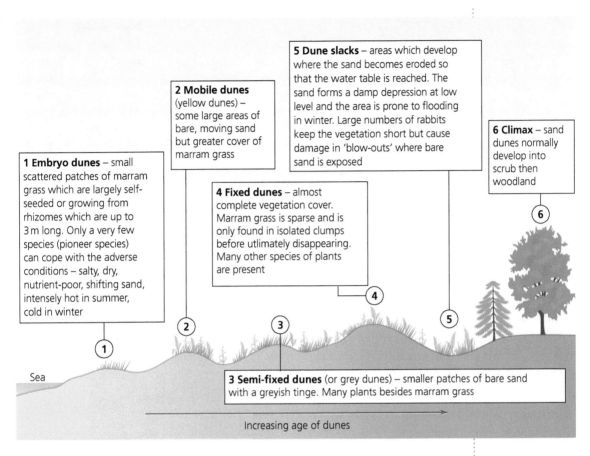

1 Embryo dunes – small scattered patches of marram grass which are largely self-seeded or growing from rhizomes which are up to 3 m long. Only a very few species (pioneer species) can cope with the adverse conditions – salty, dry, nutrient-poor, shifting sand, intensely hot in summer, cold in winter

2 Mobile dunes (yellow dunes) – some large areas of bare, moving sand but greater cover of marram grass

5 Dune slacks – areas which develop where the sand becomes eroded so that the water table is reached. The sand forms a damp depression at low level and the area is prone to flooding in winter. Large numbers of rabbits keep the vegetation short but cause damage in 'blow-outs' where bare sand is exposed

6 Climax – sand dunes normally develop into scrub then woodland

4 Fixed dunes – almost complete vegetation cover. Marram grass is sparse and is only found in isolated clumps before ultimately disappearing. Many other species of plants are present

3 Semi-fixed dunes (or grey dunes) – smaller patches of bare sand with a greyish tinge. Many plants besides marram grass

Sea

Increasing age of dunes

Figure 9 Plant succession on the Sefton coast sand dunes (a psammosere)

and chemical conditions, with a particular microclimate and soil quality. Moving from the pioneer zone (the embryo dunes) inland the following changes occur:

- physical and chemical — increasing soil moisture/decreasing infiltration rates; decreasing alkalinity of soil; increase in humus, organic content and depth of soil
- microclimate — increasing height of vegetation, leading to diminishing wind speeds and increasing temperatures
- vegetation coverage — decreasing percentage of bare ground; changes in vegetation coverage; increasing numbers of species (diversity); increasing vegetation height/layers

The improving quality of soil (increased moisture, humus and depth) creates better conditions for growth and allows larger and more diverse plants to become established. Increasing shelter and stability promote a wider range of vegetation, with more complex layers and greater diversity. The development of a succession from the initial bare surface to climatic climax community is an example of primary succession (**prisere**).

Subclimax communities

In cases where a prisere cannot develop into a climatic climax community because of arresting physical factors, such as the underlying geology producing poor soils, steep relief or poor drainage, the arrested community is called a subclimax community (Figure 10). If the arresting factor is removed, the vegetation can develop to the climatic climax community.

Exam tip

Consider the synoptic links between local-scale ecosystems and coastal landscapes. You will have studied the action of wind in coastal environments and associated landforms of sand dunes (psammoseres), as well as the action of fluvial processes in estuarine environments and the associated landforms of salt marshes (haloseres).

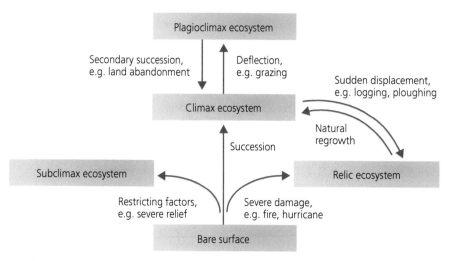

Figure 10 Generalised ecosystem evolution pathways and the effects of human impacts

Exam tip

If an exam question asks for an evaluation of the relative roles of physical and human factors in influencing succession, use variations over space and time as criteria. Ecosystems at greater distances from human influences are more likely to be affected by physical rather than human factors. Over time, human factors have assumed greater importance.

Plagioclimax communities

Repeated disturbances due to human activity can deflect succession, preventing it from reaching its climatic climax community. Human factors including fire clearance, deforestation, grazing and agriculture that are maintained over time result in a plagioclimax ecosystem.

In cases where plant colonisation can restart in an uninterrupted way following an event (such as a flood, climate change, fire or deforestation) with a higher level of plant organisation, succession is referred to as secondary succession (Figure 10).

The Arctic tundra biome
Characteristics of the climate, plants, animals and soils

The high latitude of Arctic **tundra** environments results in very low temperatures, with average temperatures ranging between −5°C and −10°C, long and dark winters during which temperatures fall below −20°C and high winds. Tundra regions are dominated by high pressure and subsiding air, leading to low mean annual precipitation of less than 150 mm. During the short growing season daylight hours are long, but the Sun's angle is so low that temperatures rarely rise above 10°C. Vegetation is dominated by mosses, lichens, grasses and dwarf shrubs. Herbivores such as caribou, musk ox and reindeer track seasonal plant growth, which insects also depend on. In summer the ground remains frozen apart from the top layers. Soils are thin, acidic and low in nutrients.

Tundra is the name given to the vast, flat, treeless Arctic region of Europe, Asia and North America in which the subsoil is permanently frozen. Tree growth is hindered by low temperatures and short growing seasons.

Knowledge check 5

Why are insolation levels low at high latitudes, and why does the seasonal amount of insolation vary so much?

Exam tip

Provide climatic data for a specific tundra location to exemplify its extreme climatic characteristics.

Climate, plant, animal and soil relationships

The growing season is extremely short in the tundra due to the low temperatures and seasonal/low-intensity insolation. Dwarf willow and stunted birch trees, with their crowns distorted by the wind, grow adjacent to seasonal rivers, but only to a maximum height of 30 cm due the short growing season, low nutrient availability and wind exposure. Growing close together helps plants to resist the effects of low temperatures and reduces wind and snow damage.

Other adaptations include the ability of plants to grow under a layer of snow, to carry out photosynthesis in extremely cold temperatures, and for flowering plants to produce flowers and seeds quickly once summer begins (Figure 11). Large numbers of insects appear, coinciding with a few weeks of high productivity. The underlying **permafrost** acts as an impermeable layer, causing waterlogging and **gleying**. Most plants therefore have short roots to avoid the permafrost and small leaves to limit transpiration. Limited plant growth results in a small amount of litter, and the lack of soil biota results in the slow decomposition of organic matter to give a thin layer of peat. Soil formation and the decomposition of organic matter are inhibited by the low temperatures. Bedrock weathered by freeze–thaw action is raised to the surface by frost heave, preventing the formation of soil horizons.

Permafrost is ground that remains at or below 0°C for at least two consecutive years.

Gleying is the process of waterlogging and reduction in soils. In waterlogged soils where water replaces air in pores, oxygen is used up by microbes feeding on soil organic matter. This results in anaerobic conditions and leads to the reduction of iron from its oxidised (ferric) state to its reduced (ferrous) state, which is soluble in the soil water. The removal of iron leaves the soil a grey/blue colour.

Knowledge check 6

What other adaptations do plants and animals have to enable them to live in tundra environments?

Figure 11 Tundra in the Canadian Arctic

Impacts of climate change

Predicted temperature increases are forecast to shift the limit of the tundra biome northwards at a rate of 15 miles per decade. The tundra is a 'stressed' biome due to its extreme climate and the short and sensitive nature of its food chain. Producers such as lichens occupy a very specific ecological niche, which is dependent on winter snow and ice. They will experience difficulties in adapting to the rate of warming. Caribou, musk ox and reindeer will need to alter their grazing ranges to keep pace with the shifting biome. Their numbers and ability to reproduce will be threatened by inadequate vegetation. In addition, higher temperatures will increase populations of biting insects, placing energy demands on these grazing mammals. Warmer winters will prevent lakes and rivers freezing, limiting movement between grazing grounds, and leading to overgrazing and the further decline of reindeer herds.

Incremental thawing of permafrost could reach a 'tipping point' of accelerated and irreversible change. Carbon- and methane-storing permafrost is now shrinking at an alarming rate as decomposition of organic matter becomes a source of heat itself, leading to a rise in soil temperatures, further decomposition and increased methane release. This positive feedback process is referred to as permafrost carbon feedback.

Exam tip

When discussing the impacts of climate change on the Arctic, consider the different specialist concepts you could apply, such as equilibrium, feedback, threshold and interdependence.

Sustainable use of the Arctic tundra biome

Threats to the Arctic tundra

Threats to the Arctic tundra include climate change, mineral exploitation and tourism. Consequences of exploiting the Arctic tundra include physical impacts, such as environmental degradation, and loss of unique flora and fauna. The most significant changes to the tundra landscape system have been caused by human activity altering the thermal equilibrium of permafrost.

Norilsk in Russia is one of the largest Arctic cities. The city is home to Norilsk Nickel, the world's biggest mining and the metallurgical complex, which contaminates the surrounding air, water and land with sulphur dioxide, nitrogen oxides, carbon monoxide, phenol, and chlorine. All vegetation is completely dead within an 8 km radius from the factory, with damaged and reduced growth as far as 200 km away.

Conflicts with indigenous populations

For centuries indigenous groups, such as the Inuit of northern Canada, Greenland and Alaska, and the Sami in Lapland, have lived in tundra regions. The cultural home of the Sami is the Sápmi region (Lapland), which stretches across the north of Norway, Sweden and Finland into Russia. In Sweden the Sami live west of the town of Őstersund in Jämtland County, where they herd and breed reindeer. Conflicts have arisen between the Sami and commercial businesses following attempts by landowners to ban reindeer herders from using their lands for grazing. This could lead to the loss of unique heritage and a culture that dates back over 1000 years.

Strategies to manage the Arctic tundra biome

Various groups are working at different levels to manage the tundra environment. At the international level there are global climate change initiatives, such as COP24 held in Poland (2018). The Wildlife Management Advisory Council (WMAC) works with the governments of the Canadian province of Yukon and the US state of Alaska, and the Canadian national government, on conserving the wildlife, habitat and traditional use of the Yukon North Slope. The Canadian government has given landmark status to Tuktoyaktuk, an Inuit town in the Northwest Territories that has 1400 **pingos**, in order to protect these landforms and the wider area from tourism.

A **pingo** is a mound 100–500 m in diameter that can rise to around 50 m above the generally flat tundra landscape, often forming the most dramatic visible landform feature.

For decades, Norilsk Nickel has ignored attempts to reduce pollution levels, despite heavy pressure from environmental groups, and has absorbed the small financial penalties. New environmental legislation from 2023 will force the company to cut its emissions by 75 per cent or risk financial penalties up to 100 times larger than now. It is anticipated that a decrease in emissions will correspond with marginal revegetation, or a likely process of secondary succession, with the expansion of tall shrubs.

> **Exam tip**
>
> You will be rewarded for providing details of strategies to manage human activity in tundra environments that are implemented at a local, national and international level.

Summary

- The distribution of the major terrestrial ecosystems of the world is principally controlled by climate.
- An ecosystem consists of biotic and abiotic components.
- Energy flows through an ecosystem and is lost at each trophic level. Nutrients are recycled within an ecosystem.
- Levels of primary productivity are limited by temperature, moisture, light and nutrient availability.
- Biodiversity is measured in different ways, and is threatened by both direct and indirect action. The biodiversity of tropical rainforests, coral reefs and wetlands is particularly at risk.
- There is a range of strategies to conserve biodiversity.
- From a 'sterile' area, a local-scale ecosystem can develop through distinct seres to form a climatic climax community.
- The climatic, biotic and soil characteristics of the Arctic tundra environment are interrelated, but are changing due to climate change.
- Human activity is increasingly threatening the fragility of the tundra environment and indigenous populations. However, strategies are being implemented to manage the Arctic tundra biome.

■ Economic growth and challenge: India

Physical background
Relief and drainage patterns

The major rivers of India originate in one of three main watersheds: the Himalaya and Karakoram ranges in the north; the Vindhya and Satpura ranges in the centre; and the Western Ghats in the west. The Himalayan river networks are snow fed and flow throughout the year. The other two networks are dependent on the monsoon and have significantly lower discharges during the dry season.

Characteristics and patterns of climate

India's climate is strongly influenced by both the Himalayas and the Thar Desert. India has six climatic subtypes, ranging from desert to alpine tundra. In general, temperatures tend to be cooler in the north, especially between September and March. India has four seasons: winter (January and February), summer (March to May), the wet **monsoon** season (June to October) and the dry monsoon season (November to December) (p. 80).

The wet monsoon season coincides with the movement of the ITCZ (p. 78) into the region, bringing an area of low pressure and drawing in hot, moist winds from the ocean. Rainfall is increased by orographic uplift (p. 79), where these moist winds are drawn over uplands such as the Western Ghats. Temperatures average 30°C and humidity is also very high, with average rainfall around 2000 mm, decreasing with distance inland. Cyclones and hurricanes are frequent towards the end of the rainy season.

The cooler dry season coincides with the extension of continental high pressure as the ITCZ moves back towards the equator and across into the tropics beyond. With high pressure dominating, there is air subsidence and out-blowing winds are dry. Temperatures remain relatively high, at 25°C, in lowland areas and evaporation rates are also high. The weather is much more severe in mountain areas.

> **Exam tip**
>
> Remember that AO2 marks are earned by making synoptic links with other parts of the A-level specification, such as the water cycle.

> **Exam tip**
>
> If one of your option themes is *Weather*, knowledge of the monsoon climate will assist with an understanding of India's physical background.

Water availability

Although precipitation is variable, India has a relatively wet climate. India ranks in the world's top ten water-rich countries, with approximately 4 per cent of world's freshwater resources. Annual precipitation provides over 4000 km³ of fresh water to India, of which approximately half is returned to oceans or evaporated, with only a small percentage stored in inland water bodies and aquifers (p. 29).

India's demographic, social and cultural characteristics

Population distribution, growth and structure

India's population — 1.3 billion in 2018 — is predicted to overtake China's by 2030 to become the world's largest, with 1.5 billion people. India's **population growth rate**, although high, has been declining, falling from 1.5 per cent in 2007 to 1.1 per cent in 2017. The southern states have lower fertility rates than the north, a contrast that will become more marked over time due to their higher education levels and successful family-planning programmes.

'**Monsoon**' is derived from the Arabic word 'mausim' meaning seasons.

> **Exam tip**
>
> Learning about low-pressure systems provides the opportunity to revisit the topics of air uplift, condensation and theories of precipitation formation covered in your study of water and carbon.

Population growth rate has two components: natural change (crude birth rate minus crude death rate) and migration.

There are considerable spatial variations in population growth rates. Kerala has a very progressive population management programme involving the education and empowerment of women, and has the lowest rate of population growth. Delhi's population growth is related to rural–urban migration fuelled by economic development and employment opportunities. The highest growth rates, in Nagaland, can be linked to the influx of refugees from Burma.

The distribution of population in India is highly uneven, with the highest concentrations in the fertile northern floodplains. Relief and climate (and associated water availability) have influenced India's population distribution. The North Indian Plains, deltas and Coastal Plains have higher population densities than the interior districts of central Indian states, the Himalayas and some of the northeastern and western states.

Although the river plains and coastal areas of India have remained regions of higher population concentration, largely due to settled agriculture, the development of irrigation (Rajasthan), availability of mineral and energy resources (Jharkhand) and the development of transport networks (peninsular states) have encouraged more recent population growth. India's urban communities have grown significantly in association with the unprecedented rate of urbanisation and industrialisation that has occurred since the economic reforms of 1991.

Exam tip

Using the correct definitions and appropriate terminology in your answers is important for strengthening AO1 marks (awarded for knowledge and understanding).

Exam tip

When assessing the relative importance of different factors influencing India's population distribution, consider the role of physical factors (relief, climate and soils). What part have historical, cultural, socio-economic and political factors played? How has the relative importance of these factors changed over time?

India has a youthful population structure. In 2018 it was estimated that 30 per cent of its population were aged under 15 years, but only 6 per cent were over 65, putting India's dependency ratio at 56 per cent, with a high youth dependency of 47 per cent and a relatively low age dependency. The dependency ratio is overstated because India has one of the highest concentrations of child workers in the world. Children mostly work in the informal sector, which is not subject to government inspections, legal protection or minimum wage requirements.

India's sex ratio is 107 males to every 100 females.

Knowledge check 7

Explain the term 'dependency ratio'.

Exam tip

You could make synoptic links between your study of India's population structure and rural–urban migration in developing countries, covered in the global governance theme. Because of youthful outmigration, the majority (around 70%) of people aged 60 years and above reside in rural areas.

Political systems and governance influencing social change

Although central government has played a role in developing India's healthcare infrastructure, each Indian state determines its own priorities. Access to healthcare is unevenly distributed between rural and urban India. Urban residents have a choice between public or private providers, but rural residents have more limited options. Public hospitals in India offer free, but less sophisticated, care than private ones. However, the system is strained to the point of collapse. Lack of access for rural communities, where two-thirds of India's population live, forces people to travel many hours to reach urban facilities. In 2018 the National Health Protection Mission was launched to improve access to primary healthcare. Higher government spending on health — only 1% of GNI at present — is important.

Compulsory education begins when children are 6 years of age and is obligatory until they are 14. While registration is compulsory, school attendance is not enforced, with only 50 per cent of children attending school. Primary school children are usually taught in the local dialect, of which there are 122 across India. Many schools educate children in Hindi, the official language of India, with English (India's second official language) as a foreign language. Education is a challenge in many rural areas, particularly the education of girls —dropout rates are high and attendance is poor. Approximately 10 per cent of young people progress to higher education. Higher education is being used as an important tool to build a knowledge-based information society.

Cultural influences, including attitudes to gender and the caste system

Gender inequality is a barrier to development. India is ranked 130th out of 189 countries in the UN's Gender Inequality Index (2017). Gender-based inequalities are evident in all aspects of Indian society (Table 1).

The **Gender Inequality Index** measures gender inequalities in three important aspects of human development: reproductive health (maternal mortality and adolescent birth rates); empowerment (proportion of parliamentary seats occupied by females and proportion of adult females with some secondary education); and economic status (female labour market participation).

Table 1 Gender inequality issues in India

Violence against women	52% of women in India think it is justifiable for a man to beat his wife, according to UNICEF
Modern slavery	An estimated 18.3 million people, mostly female, were subject to modern slavery in India in 2016
Property ownership	Women have very few land and property ownership rights
Employment opportunity	Women have limited access to employment opportunities, especially among the rural poor
Workplace discrimination	Discrimination in the workplace is common, and maternity benefits are denied by many employers
Political participation	Women are under-represented at all levels of government — national, provincial and local
Access to healthcare	Women have unequal access to healthcare, although initiatives are in place to improve the situation
Access to education	Nationally, 70% of girls attend primary school, but the figure is much lower at secondary level

Gender discrimination in India is the result of deeply entrenched norms, which favour men. Change is occurring slowly. Interventions include strengthening and increasing laws and human rights treaties, the work of NGOs, improved education and the influence of the media.

The caste system in India can also be viewed as a significant additional, cultural constraint. Hinduism is deeply rooted in India's culture, particularly through the caste system, which discriminates against the lowest caste — the Dalits or 'Untouchables'. The Indian government has tried to reduce discrimination by ensuring that a percentage of public-sector jobs is reserved for certain Dalit sub-castes. These groups need special attention because not only do they suffer from poverty and lower access to services, but they also account for the worst health outcomes in the country.

India's physical environment — opportunities and constraints

India's resource base

India's geological and metallurgical background is similar to that of mineral-rich Australia, South Africa and South America, all of which formed a continuous landmass before the break up of Gondwanaland. India's major mineral resources include coal (India has the fourth-largest reserves in the world — Table 4, p. 61), iron ore, manganese, mica, bauxite, titanium ore, natural gas, diamonds, petroleum, limestone and thorium (the world's largest deposits are in Kerala). Oilfields off Mumbai and onshore in Assam meet 20 per cent of the country's demand; however, India is still heavily dependent on imports of both coal and oil for the rest of its energy needs.

India's physical features

The Indian Himalayan region extends from the foothills in the south to the Tibetan plateau in the north. The region occupies a strategic position and borders seven countries. The north Indian states benefit from scenic, mountainous terrain, with dense forests and perennial water sources making it attractive for tourism. The fold mountains form an arc, preventing cold, Arctic winds from reaching the tropical landmass to the south. However, this region is also prone natural hazards, particularly flooding and earthquakes, and infrastructure is poor due to the steep terrain.

The Indo-Gangetic Plain lies between the Himalayas and the Peninsular Plateau. The three major river systems of India (Indus, Ganges and Brahmaputra), fed by seasonal meltwater runoff from major glaciers, have deposited alluvium across the plain. This region supports fertile agriculture but is subject to flooding. Climate change will increase meltwater discharges in the short-term but may lead to dangerous water shortages in the future. The Thar Desert forms an important southern extension of the Indo-Gangetic Plain. With rainfall as low as 150 mm per annum, this region is too arid to support agriculture without irrigation, but has huge potential for solar energy.

The Peninsular Plateau is further divided into the Central Highlands and the Deccan Plateau, located between the Western Ghats and the Eastern Ghats. The climate in

Knowledge check 8

What is the link between attitudes to gender and India's population structure? What problems are associated with India's population structure?

Knowledge check 9

Why is India's dependence on coal as its main energy source unsustainable?

Knowledge check 10

Which type of tectonic plate boundary is represented by the Himalayas?

Exam tip

If one of your options is *Energy challenges and dilemmas*, make the synoptic link between India's solar capacity and the need for alternative energy sources to reduce the country's dependence on fossil fuels, especially coal. This topic also links with the subject of India's energy security (pp. 30, 32).

the northern areas is much drier than that of the adjacent Coastal Plains. Other areas of the plateau, however, have distinct wet and dry seasons, with densely populated river valleys associated with ample access to water.

The Coastal Plains occupy a narrow strip from the Bay of Bengal in the east to the Arabian Sea to the west. Although India's coastline lacks good anchorage, given its length, important ports, such as Mumbai and Cochin (located on the Arabian Sea coast) and Chennai (on the Bay of Bengal coast), promote trade.

Constraining effects of climate variability

Human-induced climate change is associated with greater climate variability in India, including a decline in monsoon rainfall since the 1950s, an increase in the frequency of heavy rainfall events and more frequent droughts. Predictions include:

- A 2°C rise in the world's average temperatures will make India's summer monsoon more unpredictable.
- An abrupt change in the monsoon could trigger more frequent droughts as well as greater flooding in large parts of India.
- Dry years are expected to be drier and wet years wetter.
- Droughts are expected to be more frequent in some areas, especially in northwestern India, Jharkhand, Orissa and Chhattisgarh.

Climate variability has heightened India's vulnerability because its economy is heavily reliant on climate-sensitive sectors such as agriculture (more than 60 per cent of India's agriculture is rain-fed) and forestry. Predicted socioeconomic impacts include:

- a decline in crop yields by the 2040s because of extreme heat (the development of drought-resistant crops can mitigate some of the negative impacts)
- falling agricultural incomes, particularly in unirrigated areas that would be hit hardest by rising temperatures and a decline in rainfall
- health impacts, including malnutrition, child stunting (projected to increase by 35 per cent by 2050), malaria (likely to spread to where colder temperatures have previously limited transmission) and increases in mortality arising from heat waves

> **Exam tip**
>
> Consider the different specialised concepts you could apply to an exam question on the constraining effects of climate variability. Where could you integrate the specialised concepts of causality, inequality, mitigation and resilience into your discussion?

India's economic and political background

Distribution of economic activity

Agriculture in India is characterised by unequal productivity across the country:

- Andhra Pradesh has a growing agricultural sector, with opportunities for improved dairy productivity and aquaculture management.
- Madhya Pradesh has India's highest agricultural growth rate and output, including high yields of wheat, pulses and dairy products, associated with major expansions in irrigation.
- Maharashtra has a coastal economic zone, which presents potential for food processing hubs.
- Punjab continues to play an important role in grain production at both state and national levels, with more than 83 per cent of the state under intensive agriculture.

> **Exam tip**
>
> It is easy to make generalisations about India, but remember that examiners are looking for detail. Learn some statistics to support your answers and note spatial variations.

- Despite extensive areas of desert and dry land farming in Rajasthan, the state is a major producer of milk, cereals, pulses and oilseeds.
- Uttar Pradesh is an important milk-producing state.

Traditional heavy industry is concentrated in the Damodar valley. The presence of coal-mining operations and the availability of iron ore prompted the establishment of many steel and power plants in the basin area.

Using China as a model, India enacted the Special Economic Zones Act in 2005. SEZs (p. 38) are distributed across India and aim to stimulate economic activity by attracting manufacturing and services through a range of economic benefits. In 2018, 223 SEZs were operational. Sector-specific SEZs in existence or under development include:

- petrochemicals — Jamnagar, Dahej, Vizag
- pharmaceuticals and biotechnology — Pune, Aurangabad, Nanded
- manufacturing, engineering and automotive — Amritsar
- information technology — Bangalore, Kolkata

Benefits of locating in SEZs include exemption from taxes and import/export duties, liberalised labour laws and free repatriation of profits. Many sectors are allowed 100 per cent foreign-owned equity in Indian-based ventures.

Influence of political systems of democracy on economic change

After independence and Partition (whereby British India separated into India and Pakistan) in 1947, India's aim was to develop economically without the participation or influence of foreign capital. Economic policies had a strong anti-export bias. Socialist governments ensured a high level of state control over key industries, which in turn led to excessive bureaucracy and very slow economic growth.

A major economic crisis in 1991 forced the governing Congress Party to borrow money from the International Monetary Fund (IMF). This opened up the economy to economic globalisation. India is now among the fastest-growing economies in the world.

In addition to national government, state governments are also important in influencing the development of industry. States vary in terms of their politics: some adopt a free-enterprise approach; others, such as Kerala, are communist, with different levels of commitment to enacting land, labour or other business-friendly reforms.

Role of government

For the agricultural sector, central government sets price controls and tariffs. It also plays a role through central promotional schemes for particular varieties of crop and agricultural product. The individual states have responsibility for implementing agriculture policies, with each state influencing decisions for the delivery of agricultural services, technology and investment.

Political tensions with neighbouring Pakistan and China are seen as a catalyst for the development of India's aeronautical, satellite and nuclear technologies. Achieving self-reliance in defence manufacturing is a key government target, with planned

Exam tip

Diagrams that are integrated into your answers and annotated carefully to meet the requirements of the question will always earn you additional credit. A question asking about the distribution of economic activity in India would lend itself to a well-annotated sketch map.

Knowledge check 11

Identify factors other than government intervention that have encouraged the development of economic activity in India.

government spending of US$130 billion on defence industries by 2022. Some industries, such as defence and aerospace, remain under state control, but many manufacturing sectors, including vehicles, consumer electronics and white goods, are now open to **foreign direct investment**.

India has serious transport issues. The road transport sector has been declared a priority and will have access to loans with favourable conditions. Currently there are significant delays in distribution and severe bottlenecks, with state and federal governments often in opposition. The National Highways Development Program plans to improve road connectivity, including that between Delhi, Mumbai, Chennai and Kolkata (the Golden Quadrilateral). The National Railway Development Program aims to reduce congestion on rail corridors through the Golden Quadrilateral and improve port connectivity.

An extensive financial and banking sector supports the rapidly expanding Indian economy. India has a wide and sophisticated banking network. The sector includes several national and state-level financial institutions, and a well-established stock market.

> **Exam tip**
>
> When evaluating factors responsible for the rapid growth of economic activity in India, consider how the roles of these factors vary by economic sector, or over space and time.

> **Exam tip**
>
> Consider India's colonial legacy and its influence on the railway network, military structure and parts of the educational system.

India's global importance

The size and structure of India's economy

After the 1991 economic liberalisation, India achieved average GNI (gross national income p. 46) growth rates of 6–7 per cent annually. In 1991, India's GNI was ranked 16th in the world; it is currently ranked sixth, pushing France into seventh place. However, due to its large population size, per capita GNI growth has been more modest (US$7,783 in 2018).

India has become an attractive location for **MNCs** to set up factories, offices and call centres. Indian companies (e.g. Tata) have established bases in other countries, increasing India's GNI. The tertiary sector dominates (59 per cent), followed by manufacturing (26 per cent) and agriculture (15 per cent).

Rapid economic growth in India is largely the result of the expansion of the service sector rather than the growth of manufacturing. The service sector includes financial services, software services, accounting services and entertainment industries such as Bollywood, the products of which are among the most widely watched films in the world.

It is predicted that by 2040 India will be the world's second-largest economy, with China in first place and the USA third. India's economy has great potential because it

Foreign direct investment (FDI) is a financial injection by an MNC/TNC into a nation's economy, either to build new facilities (factories, shops) or to acquire or merge with an existing firm.

Knowledge check 12

In what ways has India's growth been different from that of other Asian NICs?

MNC stands for multinational corporation or company, otherwise known as a TNC (transnational company) or MNE (multinational enterprise).

Knowledge check 13

Explain the term 'positive demographic dividend'.

benefits from a large, youthful population, giving it a positive demographic dividend. English is widely spoken and graduate education is widespread. It has sophisticated space and missile technology and is a global leader in ICT.

The global shift, outsourcing and offshoring

As noted earlier, the economic crisis in 1991 triggered a major change in the economy, allowing foreign direct investments into the country, which opened India up to economic **globalisation**.

Many US and UK businesses have outsourced call centres and back-office jobs to India. Large, independent operators conduct contract work for a range of companies, from travel companies to credit card providers. MNCs such as Dell, Intel and Yahoo have built their own call centres (Figure 12). India's role as the world's outsourcing capital is the result of:

- ICT skills shortages in some developed countries
- changing technology, particularly in computers and communications. (Bangalore is a long-established technology hub, with unusually high broadband capacity. Both domestic companies (e.g. Infosys) and MNCs (e.g. Texas Instruments) have invested there.)
- India's lower labour costs
- a large English-speaking workforce. (Many Indian citizens are fluent English speakers as a legacy of British rule, giving India a comparative advantage when marketing call centre services to the English-speaking world.)
- a highly skilled and educated workforce

Globalisation is a generic term for the process of international integration in the realms of trade, economic relations and finance.

Figure 12 A call centre in India

Knowledge check 14

Give an example of an Indian MNC in the service sector.

Exam tip

Which factor is the most important in developing India's role as a global outsourcing capital?

Use of political (soft) power in the wider world

In terms of **soft power**:

- India has considerable influence over world trade as a founder signatory of the General Agreement on Tariffs and Trade (GATT), the forerunner of the WTO. India leads the developing nations in global trade negotiations and is trying to encourage a more liberal global trade regime, especially in services.
- India's political influence in global organisations, such as the World Bank and International Monetary Fund, is increasing. India therefore has a growing influence over global financial decision making.
- Indian MNCs such as Tata Steel and JSW have an important economic influence within the global community.
- the five permanent members of the United Nations Security Council support India's proposal to join them, as a result of India being one of the most consistent contributors to UN peacekeeping operations. India is also playing a greater role in disaster response.
- India is committed to democratic institutions, the rule of law and human rights, has a huge and talented diaspora, shares many western values and is culturally rich.

> **Soft power** refers to the use of non-coercive power (cultural, institutional and ideological) to cement a country's leadership position in the world. Hard power includes the use of military force, sanctions and trade barriers.

Threats to India's environment

Environmental pressures

Economic growth is associated with industrial pollution. The Ganga (Ganges) River, which has great spiritual and emotional significance for Indians, is seriously polluted. 764 industrial units along the main stretch of the River Ganges and its tributaries discharge 500 million litres of toxic waste a day.

Agricultural output is linked to soil erosion, with 30 per cent of India's gross agricultural output lost every year to soil degradation, poor land management and counter-productive irrigation.

After independence India's forests were exploited commercially for pulp and paper. India is witnessing a rising demand for forest-based products, leading to deforestation and encroachment into protected forest areas. Although the overall forested area is increasing, the loss of biodiversity is not sustainable.

Nearly 30 per cent of India is facing **desertification** due to wind and water erosion, vegetation degradation, salinity and the growth of settlements linked to increased economic activity.

> **Desertification** is the process by which once-productive land gradually changes into a desert-like landscape. It usually takes place in semi-arid areas on the edges of existing desert. The process is not necessarily irreversible.

> **Exam tip**
>
> You could make synoptic links between the environmental pressures associated with economic activity (e.g. deforestation) and changes to both water and carbon cycle stores and flows. Consider variations over temporal (time) and spatial scales of these stores and flows.

Water security, food security and energy security

Water

India struggles with both too much and too little water. The areas that flood in the monsoons, and the number of people that are affected, increase every year. Water demand is high due to rapid urbanisation, rising standards of living, industrialisation and extensive agriculture. The unpredictability of rainfall in many semi-arid areas is increasing because of failed monsoons and climate change.

Physical constraints, distribution patterns, technical limitations and poor management prevent India from accessing its water resources effectively. India is designated a region of 'water stress', with utilisable fresh water below the international standard limit of $1700\,m^3$ per capita per year. In future it may become designated a 'water scarce' region if this falls below $1000\,m^3$ per capita per year. Fifteen per cent of India's groundwater resources are overexploited. The efficient use of ground water resources will need to be incentivised to offset further falls in water table levels.

Food

India's growing population increases the challenge of **food security**. Despite rapid economic growth in India, millions of Indians suffer poverty and hunger. Previously self-sufficient in wheat, India now imports grain to feed the rapidly expanding population, which is increasing by 17 million each year. There is growing demand for fertile farmland to be used by MNCs to grow industrial and food crops for export. These developments also use up water resources and increase pollution of soil and water.

Poor farmers are often forced onto more marginal land that, without expensive fertilisers and pesticides, produces lower yields. Those farmers who do try new technologies are at risk of debt if crops should fail (Figure 13).

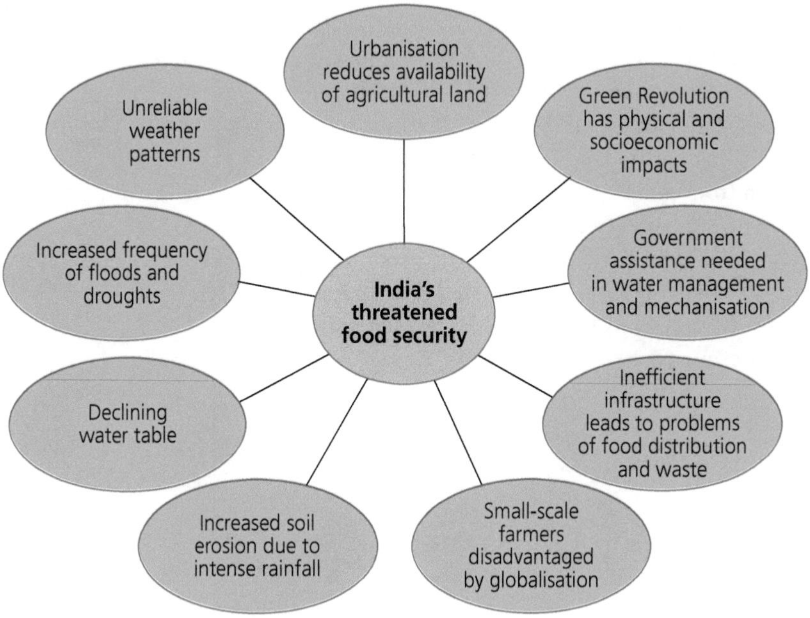

According to the UN Food and Agriculture Organization (FAO), '**food security** exists when all people, at all times, have physical and economic access to sufficient, safe and nutritious food that meets their dietary needs and food preferences for a healthy life'. The FAO identifies four dimensions of food security: the physical availability of food; economic and physical access to food; food utilisation (includes feeding practices and diversity of diet); and stability.

Figure 13 India's food security under threat

Exam tip

In your answers on this topic you should emphasise the interdependence between food and water security.

Energy

Current peak electricity demand is around 135 GW, but that is projected to more than double by 2022 to 283 GW. Beneath these figures lies a huge suppressed demand: the World Bank estimates that as many as 400 million people in India currently have no access to electricity. India relies heavily on fossil fuels, especially coal. Nuclear contributes 4.8 GW and large hydropower (HEP) plants 40 GW. The proportion contributed by other renewables is significant, but small, at 29 GW. Most of that (70 per cent) is from wind.

The growth rate in India's energy consumption is one of the fastest of all the major economies. There are significant reserves of coal, but limited oil and gas resources. At present coal is the dominant energy source, accounting for 56 per cent of India's total electricity generation (2018). The burning of coal produces greenhouse gases, sulphur dioxide, nitrogen oxides and particulate matter. Although the proportion of coal-derived energy is falling, with 550 thermal power station projects cancelled in the past 7 years, coal will continue to be significant for several decades to come.

Rapid urbanisation

It took nearly 40 years (from 1971 to 2008) for India's urban population to rise by nearly 230 million; it will take only half that time to add the next 250 million. For the first time in India's history, five of its largest states have more of their population living in cities than in villages. This rate of growth has produced a crisis in urban environments, characterised by water pollution from untreated sewage and air pollution from industry and car exhausts. The numbers of road vehicles and construction sites have multiplied, and outdoor air pollution has become a major health hazard and killer. This adds to the already large burden of ill-health caused by household air pollution from the use of solid fuels for cooking.

Mumbai is classified as a **global city**. Four other Indian cities were ranked in the top 100 global cities in 2016: Chennai, Delhi, Kolkata and Pune. In 2015 Mumbai's urban area registered a population of 22 million, more than double that of 1970. Migration has been from the impoverished states of Uttar Pradesh and Bihar. Mumbai has witnessed a more than four-fold rise in vehicular traffic since 2001. Water pollution from untreated sewage and air pollution from industry and car exhausts present major challenges. Waste plastic littering Mumbai's streets and choking its rivers has become a serious environmental and health hazard. High-intensity monsoon rainfall causes flooding due to sewer failures. Its low-lying densely populated coastline is also threatened by a future rise in sea levels. Mumbai has the world's largest population exposed to coastal flooding and sea water intrusion.

A **global city** (also called world city or sometimes alpha city) is one that is deemed to be an important node in the global economic system.

> **Exam tip**
>
> In any discussion of environmental pressures associated with rapid urbanisation, provide place-specific detail and avoid restricting your answer to only one city.

Sustainable development in India

Strategies to manage environmental problems associated with economic growth

Every year, India's central government distributes tax revenue to states, based on a formula that includes each state's population, area and poverty levels. In 2014, forest cover was added to the formula, creating an important new financial incentive for state governments to protect and restore forests. The government estimated that, from 2015–2020, it would devolve US$6.9–12 billion per year to states, based on their forest cover in 2013. States that increase forest cover will stand to gain tax revenue, while states that lose forest cover will lose it. Each state's forest coverage is detected by biennial satellite monitoring carried out by the India Forests Survey.

India's reform is an example of an ecological fiscal transfer (EFT), in which higher levels of government devolve funds to lower levels of government based on their success in achieving environmental indicators. EFTs, together with REDD+ (see below) and payments for ecosystem services (PES), encourage the protection and restoration of forests.

> **Exam tip**
>
> If one of your chosen themes is *Ecosystems*, knowledge of payments for ecosystem services (PES) reinforces your understanding of the value of ecosystems as providers of goods and services.

REDD+ Strategy India, one of the projects aimed at supporting India's commitment to the Paris Climate Agreement, was launched in 2018 by the Minister for Environment, Forest and Climate Change. REDD+ stands for 'Reducing Emissions from Deforestation and forest Degradation' and aims to achieve climate change mitigation by incentivising forest conservation.

Strategies to improve water security

Cities that suffer from both excess water during the monsoon season and periods of drought, such as Chennai, have attempted to improve their water security through:

- rainwater harvesting — 75 per cent of homes have systems for harvesting rainwater
- artificially recharging groundwater
- recycling sewage
- desalinisation — the Nemmeli Desalinisation Plant, one of Chennai's four desalination plants, located 35 km south of the city on the Bay of Bengal, cost over 5000 million Indian rupees

To maintain a sustainable water supply in the future, cities such as Chennai need to:

- increase water supply from non-rainfall-dependent sources (recycling sewage and desalination)
- maximise storage of monsoon rainfall
- reforest river catchments to improve groundwater supplies
- use reverse osmosis for the treatment of polluted groundwater and recycling of wastewater
- introduce demand management solutions, such as water conservation and the reduction of leakages

Sustainable development is development that meets the needs of the present, without compromising the ability of future generations to meet their own needs.

Desalinisation is the process by which dissolved solids in sea water are partially or completely removed, to make it suitable for human use.

Knowledge check 15

Why is desalinisation a limited option for increasing India's water supplies?

Strategies to improve food security

In the 1960s HYVs (high-yielding varieties) were introduced to India as part of the **Green Revolution**, and had a major impact on Indian food production. The Green Revolution was successful in raising incomes for farmers on naturally fertile soil, but increased inequalities between wealthy farmers on productive land and poorer farmers on marginal land.

The Indian government introduced the National Food Security Act in 2013. Its objective is to provide food security by ensuring that up to 75 per cent of the rural population and 50 per cent of the urban population receive 5 kg of food grain per person per month at subsidised prices. However, this strategy does not address the issue of improving the supply of food.

To maintain a sustainable food supply in the future India needs to:

- increase government investment in agriculture (machinery, seeds and agrochemicals)
- improve prices for farmers
- reduce input costs
- improve local food security through food distribution centres, as current storage and distribution systems are inefficient
- improve the management of water supplies

> **Green Revolution** refers to the application of modern farming techniques, including a package of technology (fertilisers, pesticides, water control and mechanisation), to developing countries. It has more recently incorporated the Gene Revolution, following the adoption of genetically modified crops.

Strategies to improve energy security

India has considerable potential for the exploitation of renewable energy resources, such as solar and wind power, and biofuels (from sugarcane). Although coal will remain the dominant fuel source, its share of generation is predicted to fall as that from renewables rises. The Desert Power India — 2050 plan, put forward by India's state-owned power utility, the Power Grid Corporation, predicts that 455 GW of electricity could come from renewable sources by 2050, and around two thirds of that would be produced by vast solar PV installations in areas such as the Thar Desert.

India's National Action Plan on Climate Change (NAPCC) aims to improve energy efficiency and develop renewables, especially solar power. A proxy carbon tax is currently levied on coal, both imported and domestic. However, because of India's economic growth, GHG emissions are predicted to be in the range of 4.0–7.3 billion tonnes by 2030, compared with 2.43 billion tonnes in 2010.

> **Energy security** is achieved when there is an uninterrupted availability of energy at a national level and at an affordable price.

> **Exam tip**
>
> Consider some of the challenges associated with the Desert Power India — 2050 plan. For example, basic infrastructure such as roads and power cables are lacking, skilled labour would have to come from elsewhere, and keeping sensitive solar PV panels free of dust and grit in a desert environment would be problematic.

Strategies to improve the sustainability of urban communities

The introduction and enforcement of anti-pollution legislation lags behind India's economic growth. Vehicle emissions need to be reduced, but there are no regulations for emissions of sulphur dioxide and nitrogen oxide. City governments in India are still focused on investing in road expansions and overpass construction projects rather than curbing car dependence and improving public transport services. As living standards increase it is anticipated that pollution controls in cities will be introduced. Some progress and proposals have been made:

- Banning diesel autos and vans (e.g. in Delhi and Imphal). In Imphal there are plans to substitute auto-rickshaws and diesel-van transport with a state bus service and e-rickshaws.
- Some Indian businesses are taking the initiative to implement transport demand management (TDM) strategies to improve the productivity of their employees and reduce the social costs of car congestion. Wipro has worked with the Bangalore Metropolitan Transport Corporation (BMTC) on specific routes designed to move workers more efficiently and reduce Wipro's employees' carbon footprint.
- In Maharashtra state, of which Mumbai is the capital, a ban on single-use plastic items has been introduced.
- Proposals include switching to clean energy sources for cooking stoves, public transport and industry, as well as measures to reduce road traffic by raising fuel taxes and parking fees, levying congestion charges, and creating vehicle-free zones and cycle paths.
- Development of rural areas could do much to stem internal migration and take pressure from urban centres.

> **Exam tip**
>
> When answering questions about sustainable development in cities avoid discussing urban problems in general and ensure that you provide precise locational information.

Summary

- The physical background of India can be divided into the Himalayan mountains to the north, the Indo-Gangetic Plain in the centre and the Peninsular Plateau to the south. India has six climatic subtypes ranging from desert to alpine tundra.
- India's population is predicted to overtake China's by 2030 to become the world's largest. The distribution of population is uneven, as is access to healthcare and education. Gender inequality and the caste system are both barriers to development.
- India has large reserves of coal but needs to import 80 per cent of its oil. The physical environment creates opportunities, such as mountains for tourism, fertile soils for agriculture and deserts for solar power generation, but also presents challenges, such as floods and earthquakes. Climate variability has heightened India's vulnerability.
- A major economic crisis in 1991 forced the Indian government to borrow money from the IMF: this opened the economy up to globalisation.
- India has become the world's outsourcing capital for services and has been assuming an increasingly important role in global economic and political systems.
- Threats to India's environment associated with economic growth include environmental pressures of fossil fuel use, industrial pollution, soil erosion, deforestation and desertification, issues of water, food and energy security, and rapid urbanisation.
- EFTs are being used to manage deforestation in India. Water security is being improved through desalination, wastewater recycling, rainwater harvesting and improved demand management. Food security has been improved through the Green and, more recently, the Gene Revolution, and through government subsidising food grains, although more progress is needed. There are plans to place renewables ahead of fossil fuels in India's energy mix.

■ Economic growth and challenge: China

Physical background

The People's Republic of China is located in east Asia, on the west side of the Pacific Ocean. It is the third-largest country in the world after the Russian Federation and Canada.

Relief and drainage patterns

China's relief is both complex and variable, ranging from mostly mountains, high plateaus and deserts in the west, to plains, deltas and hills in the east. The Qinling Mountains provide a natural boundary between north and south China.

The Tibetan Plateau in the west is the source of almost 50 per cent of the major river systems in China, including the three longest rivers: the Yangtze, Huang He (Yellow) and Pearl. These flow west to east, into the Pacific Ocean. About 10 per cent of Chinese river systems drain into the Indian or Arctic oceans. The remaining 40 per cent have no outlet to the sea; they drain through the dry western and northern areas of China, forming deep underground water reserves.

Characteristics and patterns of climate

Although most of China lies in the temperate belt, its climatic patterns are complex, ranging from subtropical in the south to sub-Arctic in the north. Monsoon winds (p. 79) dominate the climate and have a major influence on the timing of the rainy season and the amount of rainfall. Alternating seasonal air-mass movements and accompanying winds produce moist summers and dry winters.

Water availability

The main problem facing China is water distribution, rather than availability. China has 7 per cent of the world's fresh water, but 80 per cent of China's water supply lies in the south. Annual per capita availability is $2300\,m^3$, but there are significant regional differences — north China's water availability is approximately $271\,m^3$ per capita. Pollution contributes to water scarcity, and demand is growing, with consumption forecast to rise to 670 billion m^3 per year by the early 2020s.

> **Knowledge check 16**
>
> What strategy is in place to transfer water to northern cities such as Beijing?

China's demographic, social and cultural characteristics

Population distribution, growth and structure

China's population of 1.4 billion is the largest in the world. It is concentrated in the eastern and coastal part of the country and along major rivers. Much of the country, including the steep Himalayas, the dry grasslands and Gobi Desert in the north, and the central region, is almost uninhabited.

In the 30 years following the 'Open Door' (p. 39) reforms of the late 1970s, 300 million people migrated from rural areas to the industrialising Chinese cities along the east coast (Figure 14). Rural–urban migration to the core regions along the coast, and increasingly along the Yangtze corridor, occurred in response to economic forces and the employment opportunities in urban areas and SEZs (p. 38). Nearly 60 per cent of Chinese now live in urban areas. A strict registration system called *hukou* has restricted further migration. The demand for labour in cities has led to rural migrants being given a temporary urban *hukou*, to allow them access to housing and basic welfare. Registration requires annual renewal, so that the authorities can control the number of migrants. Temporary *hukous* set the migrant population apart, and has increased inequality in terms of access to services in cities.

Hukou is the Chinese registration system that officially recognises an individual as a resident of an area.

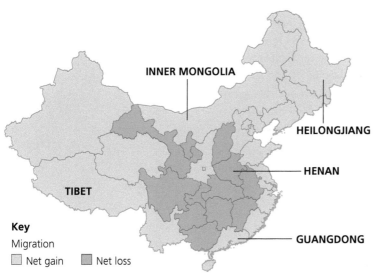

Key

Migration

☐ Net gain   ▨ Net loss

- Rural Henan province had the biggest exodus of 10.25 million (more than the entire population of Sweden)

- Urban Guangdong province received 20.5 million migrants (more than the entire population of Romania)

Figure 14 Rural–urban migration and the redistribution of China's population, 1978–2010

Exam tip

Note that economic factors, such as the 'Open Door Policy', are driven by the Communist party and therefore it is difficult to disaggregate economic and political influences. These two variables are often interdependent in China.

Exam tip

When assessing the relative importance of different factors influencing China's population distribution, consider the role of physical factors (relief, climate and soils). What role have political, cultural and socioeconomic factors played?

China's population growth rate (p. 20) continues to decline, falling from 0.9 per cent in 2000 to 0.4 per cent in 2017. Under Mao Tse-Tung (Mao Zedong), China nearly doubled in population from 540 million in 1949 to 969 million in 1979. This growth slowed because of the One-Child Policy instituted in 1979. Although the policy was lifted in 2015, birth rates remain low due to socioeconomic changes that discourage second children, such as the high cost of education and a shortage of affordable childcare. The highest population growth rates occur in ethnic minority areas, such as the autonomous regions of Ningxia and Xinjiang, where the One-Child Policy was less strictly applied.

China has an ageing population structure. 2018 estimates are that 18 per cent of its population is aged under 15 years, with 9 per cent over 65. This puts China's dependency ratio at 37 per cent, with a youth dependency of 25 per cent and a relatively high age dependency of 12 per cent. China's population is growing old at a faster rate than almost all other countries. Its sex ratio is 108 males to every 100 females.

Political systems and governance influencing social change

China's centralised healthcare system has undergone several transitions:

- The introduction of 'barefoot doctors' (peasants who received short, intensive medical training, providing virtually free basic medical services in rural China) in the 1960s resulted in an increase in life expectancy and a fall in infant mortality.
- Barefoot doctors were replaced by a market-based healthcare system during the economic reforms of the late 1970s and 1980s. However, the system was limited by financial constraints and ultimately resulted in declining health coverage.

Approximately 95 per cent of the population is now covered by universal healthcare, with China having a lower uninsured rate than the USA. However, there is a marked difference between urban and rural areas, with more than double the number of hospital and health centre beds per thousand people in urban compared with rural areas.

By law, each child must have 9 years of compulsory education from primary school (6 years) to junior secondary education (3 years). Educational access remains uneven in China. Children from affluent, urban families generally have greater access to high-quality education than those from lower-income backgrounds. The household registration system (*hukou*) has further widened this gap. Government policies that require local governments to provide some of the funding for schools have compounded these inequalities, as less affluent areas with insufficient resources to pay skilled teachers, purchase materials and maintain school facilities record lower levels of educational attainment.

Cultural influences, including attitudes to gender and minorities

There are officially 56 ethnic groups in China, but 92 per cent of the population belongs to one group — the Han Chinese. The most ethnically diverse areas are in west China, where 80 per cent of the country's minorities live. China is facing growing criticism over its persecution of some minority groups, holding up to one million Uighur Muslims and other Muslim groups in internment camps in western Xinjiang.

Out of China's population of 1.4 billion, there are nearly 34 million more males than females — the equivalent of almost the entire population of Poland. China's One-Child Policy was a huge factor in creating this imbalance, because of millions of couples' preference for a son. Problems associated with this gender imbalance include rising crime rates and the development and growth of 'bachelor villages'.

> **Exam tip**
>
> You could make synoptic links between your study of China's population structure and rural–urban migration in developing countries, covered in the *Global governance* theme. Youthful outmigration has left rural areas with an ageing workforce. This threatens future agricultural production and China's food security.

China's physical environment — opportunities and constraints

China's resource base

China has a range of natural mineral resources including coal, iron ore, petroleum, natural gas, mercury, tin, tungsten, antimony, manganese, molybdenum, vanadium, magnetite, aluminium, lead, zinc and uranium. Despite China's relatively rich endowment in coal, its low quality and uneven distribution constrains China's economic development. The greenhouse gas emissions and pollution associated with the combustion of low-grade coal has led to increases in the cost of environmental amenity and repair, placing a strain on China's economy. Many important metallic minerals such as iron, manganese, aluminium and copper are of poor quality and difficult to smelt. However, China has the world's largest hydropower potential. In 2017, the country added 9.12 GW of installed hydropower capacity, bringing its total to 341 GW.

Knowledge check 17

Why is China's dependence on coal as its main energy source unsustainable?

China's physical background

Almost two-thirds of China is mountainous, comprising rugged plateaus, foothills and mountains, with these areas of higher altitude concentrated in the west. The relief can be divided into:

- the Qinghai-Tibet Plateau (mean height over 4000 metres), the highest and largest plateau in the world. Although remote, accessibility has recently been improved by the construction of the Qinghai–Tibet Railway, which has opened up the region to tourism.
- large basins and plateaus at altitudes of 1000–2000 metres. These include the Inner Mongolian and Loess Plateaus and the Sichuan Basin. Hazards include sandstorms on the Loess Plateau and earthquakes in Sichuan.
- the plains of northeast and northern China, and the middle-lower Yangtze at altitudes of over 500 metres. These plains are fertile, well-cultivated and support high population densities.

China's varied landscape, including the picturesque karst landscapes in Guilin, lakes in Jiuzhaigou and the Rainbow Mountains in Zhangye, present many opportunities for the development of tourism.

In addition to facing threats from flooding, droughts (see below) and sandstorms, China is located in an active seismic zone and can experience major earthquakes. The 2008 Sichuan earthquake led to a death toll of 69,000. The typhoon season in China normally runs from May to November, affecting southern and eastern coastal regions.

Exam tip

If one of your optional units is *Glaciated landscapes*, make a synoptic link with the geomorphological conditions under which loess forms.

Knowledge check 18

Use your knowledge of tectonic plate boundaries to explain the cause of seismic activity in China.

Constraining effects of climate variability

China's mean annual air temperature has increased by more than 1.0°C in the past three decades — higher than the global average. Although its annual precipitation has not changed notably, its regional and seasonal patterns have changed significantly. China faces a major threat from flooding (in 2013, more than 230 Chinese cities experienced floods); by contrast, more than half of China's cities face the threat of water scarcity.

Rapid urbanisation means that many of China's infrastructure networks (among the world's largest) are increasingly vulnerable to flooding and drought. Nearly 80 million Chinese city dwellers live in coastal zones at risk of sea-level rise (this compares with 30 million in India and 20 million in the USA). The frequency and intensity of extreme droughts are projected to increase significantly under future climate change. Water scarcity could hamper many of China's most important policy objectives, including continuing urban growth and achieving food self-sufficiency.

China's economic and political background

Distribution of economic activity

Since 1979, six **special economic zones** (SEZs) and 14 open cities have been established (Figure 15). These offer reduced restrictions on land, labour, wages, taxes and planning regulations to overseas firms, especially those involved in high-technology industries. The result has been the emergence and dominance of economic activity in coastal areas, which have received most of the internal investment as well as having imported capital, technology and entrepreneurial skills, at the expense of the interior. The growth in economic activity in these coastal locations, which offer minimal costs and maximum export opportunities, has been reinforced by the investment decisions of MNCs, high levels of rural–urban migration and infrastructure improvements.

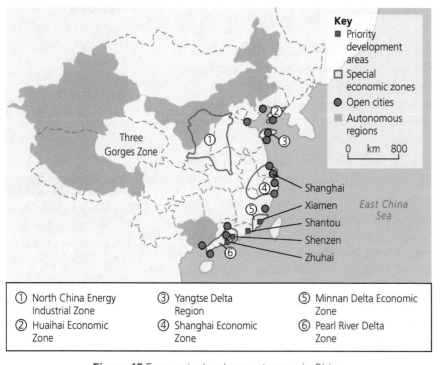

① North China Energy Industrial Zone	③ Yangtse Delta Region	⑤ Minnan Delta Economic Zone
② Huaihai Economic Zone	④ Shanghai Economic Zone	⑥ Pearl River Delta Zone

Figure 15 Economic development areas in China

> **Exam tip**
>
> Consider the different specialised concepts you could apply to an exam question on the constraining effects of climate variability. Where could you integrate the specialised concepts of causality, inequality, mitigation and resilience into your discussion?
>
> ....
>
> **Special economic zones** (SEZS) are industrial areas where favourable conditions are created to attract foreign MNCs. Conditions include low tax rates and exemptions from tariffs and export duties.

> **Knowledge check 19**
>
> Why has economic growth in China been concentrated in coastal areas?
>
> ....

In order to address the regional inequalities created by the concentration of economic activity along China's east coast, the government has been instrumental in implementing the West China Development project, created in 2000, to help the western provinces catch up with coastal areas. Raw materials, such as coal, also influence the location of manufacturing industry, with most deposits concentrated in the northern part of the country. Agricultural activity is concentrated on the fertile, low-lying plains and deltas to the east.

> **Exam tip**
>
> Although FDI (p. 26) has been essential for China's economic growth, you should be aware of the recent growth and contribution of Chinese MNCs, as well as domestic industrial developments and markets.

Influence of political systems of modified communism on economic change

Between 1949 and 1976 government locational decisions were dominated by Marxism-Leninism, with a socialist, collectivist and centrally planned agenda. After the death of Mao Tse-Tung (Mao Zedong) in 1976, China's economy took a major change in direction. In 1978, Deng Xiaoping, the new leader of the Chinese Communist Party, introduced the 'Open Door Policy', which was designed to overcome China's isolation from the world's economies. The country had become increasingly aware that the world, and southeast Asia in particular, was developing and leaving China behind. The development of economic activity in China includes the growth of manufacturing, service and financial industries, as well as agriculture. Rapid economic growth in China, initially due to the expansion of manufacturing, is increasingly driven by the service sector.

Knowledge check 20

Name a Chinese MNC.

Role of government

In 1978, three quarters of the country's industrial production was accounted for by centrally controlled, state-owned enterprises (SOEs), following centrally planned output targets. The 1980s focus on increased productivity forced SOEs towards reform. Large SOEs improved their management and smaller SOEs eventually privatised. SOEs have attracted MNCs as partners and FDI (foreign direct investment) has been highly significant. FDI in China increased from US$3.5 billion in 1990 to US$139 billion in 2016. The 'Open Door' reforms allowed China to embrace globalisation whilst remaining under one-party, communist rule. China became the world's manufacturing hub, specialising in the labour-intensive, export-led production of cheap goods, which enabled a gradual increase in product complexity. Although established by a communist government, SEZs were deliberately located in coastal areas far from the centre of political power in Beijing, minimising political influences.

During Mao's era, rural industries known as town and village enterprises (TVEs) produced heavy goods such as iron, steel, cement, chemical fertiliser and farm tools, as well as hydroelectric power. After 1978, these enterprises expanded to develop a wider range of businesses. TVEs became the backbone of development in rural areas, playing a catalytic role in transforming the Chinese economy from a command economy to a market economy.

Knowledge check 21

Identify factors other than government intervention that have encouraged the development of economic activity in China.

Under Mao, collectivised agriculture was the norm, leading to agricultural stagnation. Some of the earliest reforms of the late 1970s occurred in rural areas. Communes were dismantled and although land remained state-owned, leases were given to households under the 'responsibility system'. Once a state quota was met, farmers could market any surplus produce. As a result, China's agricultural base strengthened, which helped to drive China's food production forward.

China's President Xi Jinping consolidated his position of power in 2017, when he was elevated to the same status as former Chairman Mao. He has pledged to address environmental concerns and corruption, whilst maintaining the politburo's grip on the economy. The 2018 trade war between the two economic powerhouses of China and the USA has placed increased pressure on China's economy.

China's global importance

The size and structure of China's economy

In 1976 China's GNI (p. 46) was ranked 124th in the world; it is currently ranked second. However, due to its large population size, per capita GNI growth has been more modest (US$8,827 per capita). China's economic boom coincided with a positive demographic dividend, where there were more people of productive age with a low dependency ratio. In 2017 China's share of the world economy was 15 per cent, compared with 24 per cent for the USA. Total trade, however, was much higher, at US$4,233 billion, compared with US$3,233 billion for the USA.

It is predicted that China's GNI will overtake that of the USA in 2029 at projected GNI growth rates of 2 per cent for the USA and 6.5 per cent for China. Its manufacturing-based economy has great potential because China benefits from a highly educated and technically innovative population, and a modern transport infrastructure. It leads the world in fields such as renewable energy, and its military technology is growing and challenging that of the USA.

China's **economic miracle** showed signs of slowing in 2015 after three decades of GNI growth rates of 10 per cent. Although China's economy is growing, it is not becoming more efficient. In 2017 primary industry accounted for 8 per cent of GNI, secondary for 40 per cent and tertiary 52 per cent, reflecting the shift to economic growth based on services and consumerism associated with a growing middle class. Weaknesses include rising wages, which are making China's economy a higher-cost location for MNCs, China's heavy reliance on imported materials and the high costs of addressing air and water pollution. The negative demographic dividend, associated with the rapidly ageing population, will soon affect China. As people live longer, they need to either accumulate wealth or face a reduction in their standard of living in their old age.

The global shift, outsourcing and offshoring

The liberalisation of trade since the 1980s has allowed global MNCs to establish branch plants or trade relationships with Chinese-owned factories in SEZs. China's economy has since matured, with companies moving further up the manufacturing value chain to produce high-value goods as a result of government strategic planning.

> **Exam tip**
>
> Evaluate the roles of the factors responsible for the rapid growth of economic activity in China. Do these vary by economic sector, or over space and time?

> An **economic miracle** is a period of unexpectedly strong and rapid economic development.

China needs resources for its continued economic growth and has been determined to establish trading relationships with countries that can supply raw materials. The increase in food and mineral imports into China has had the effect of driving up many world commodity prices, such as those for iron and other ores. China has increased trade and FDI on all continents, but there have been periodic rows with China's trading partners concerned about the 'dumping' of exports. China has always traded with Africa, but attention is increasingly being focused on that trade and the political influence that accompanies it.

Use of political (soft) power in the wider world

In 2014 Chinese overseas investment surpassed FDI into China. China was home to 172 of the world's top 2000 companies in 2016, with the greatest representation of any Asian country, exceeding Japan's 127, South Korea's 67 and India's 56 companies.

China's political global influence has grown. Chinese strategic objectives include defending national sovereignty and territorial integrity, acquiring regional pre-eminence and safeguarding overseas interests. It has military ambitions to build a **blue water navy**. China is, in the short term, attempting to replace the USA as the dominant power in east Asia and, in the long term, to challenge the USA's position as the dominant power in the world. China is an active participant in global organisations, governance, conventions and treaties, such as the IMF, WTO, World Bank, UN and IPCC.

China's political influence is most strongly felt in east Asia, which has become increasingly China-centric. China has successfully managed commercial, political and military agreements with states of the ASEAN community, and others in Africa and South America. China's ambitions extend to space, with its space agency achieving the first soft landing on the far side of the Moon in early 2019, China's second Moon landing overall.

> **Exam tip**
>
> Many of China's TNCs are state owned; therefore, it is difficult to disaggregate China's global economic and political importance. China's global 500 companies (MNCs) are mostly state-owned, for example Sinopec.

Threats to China's environment

Environmental pressures

Environmental pressures associated with China's rapid economic growth include air and water pollution, soil erosion, deforestation and desertification.

Despite attempts by central government to reduce China's reliance on coal, there was a surge in new coal projects approved at provincial level between 2014 and 2016. This happened because of a decentralisation programme that shifted authority over coal plant construction approvals to local authorities. China's heavy reliance on fossil fuels has resulted in increased emissions of carbon dioxide, nitrous oxides, acid precipitation (which falls on 30 per cent of China) and smog. Air pollution is a major issue in cities such as Beijing and Shanghai due to their heavy reliance on coal.

Knowledge check 22

What are the costs and benefits of Chinese investment in Africa?

A **blue water navy** is one that operates in the open ocean, whereas a green water navy has ships that can only operate close to the coast.

Approximately one-third of industrial wastewater is released into rivers and lakes without being treated. As a result, 90 per cent of urban waterways and lakes are severely polluted and major pollution incidents are common. Underground water supplies in 90 per cent of China's cities are contaminated.

Illegal logging and slash-and-burn agriculture remove up to $5000\,km^2$ of forest each year. In northern and central China forest cover has been reduced by half in the last two decades. Southern provinces are threatened by plantations: in Hainan and Yunnan, for example, indigenous trees are felled to make way for fast-growing eucalyptus plantations for paper pulp. Up to 400 million people are at risk of desertification in China, with the affected area covering a third of the total land area. Much of it is happening on the edge of the settled area, which suggests that human activities of poor land use, overcultivation and overgrazing are largely to blame. Deforestation and desertification are linked and result in excess runoff and associated soil erosion. It is estimated that 40 per cent of China's territory — over 3 million km^2 — suffers from soil erosion.

> **Exam tip**
>
> Make the synoptic link between the environmental pressures associated with economic activity and changes to both water and carbon cycle stores and flows. Consider variations in the temporal (time) and spatial scales of these stores and flows.

Water security, food security and energy security

Water

Water resources in China of over $2000\,m^3$ per person are above the level where **water stress** starts. However, water resources are unevenly distributed, meaning that eight northern provinces suffer from **acute water scarcity** and four from **water scarcity**, while a further two (Xinjiang and Inner Mongolia) are largely desert. These eight provinces account for 38 per cent of China's agriculture, 46 per cent of its industry, 50 per cent of its power generation (coal and nuclear use a lot of water), and 41 per cent of its population.

Water pollution is a major source of health problems in China due to untreated waste products. One-third of all rivers, 75 per cent of major lakes and 25 per cent of coastal zones are currently classified as highly polluted.

Food

There are considerable spatial variations in food production across China, resulting in regions of food surplus and of food deficit. Nine provinces (including Tibet and Yunnan) out of 31 have been classified as food insecure. In these provinces 60 per cent of the population consume less than the recommended target of grain products.

As dietary consumption patterns become more westernised, Chinese food consumption will have an increasing transnational footprint, with impacts on agroecological systems elsewhere. The IPCC (International Panel on Climate Change) predicts lower yields of certain crops in parts of China associated with climate change.

> **Water stress** occurs when annual water supplies drop below $1700\,m^3$ per person.
>
> **Water scarcity** occurs when annual water supplies drop below $1000\,m^3$ per person.
>
> **Acute water scarcity** occurs when annual water supplies drop below $500\,m^3$ per person.

> **Exam tip**
>
> In your answers you should emphasise the interdependence between food and water security.

Energy

China is the world's largest consumer of energy. Coal accounted for almost two-thirds of China's primary energy consumption in 2016. Oil is the second-largest energy source, but 65 per cent of consumption in 2016 had to be imported, with the percentage expected to rise to 75 per cent by 2035. While not as stark as its dependency on imported oil, China's reliance on imported natural gas is also significant. The government has made mitigation of climate change a top priority, and therefore needs to transition its energy mix away from fossil fuels.

Rapid urbanisation

China has more than 100 cities with a population greater than 1 million and six of the world's **megacities**. Millions of people have migrated from rural to urban areas to fill the jobs generated by the economic explosion. Most migrants head for the eastern-seaboard cities of Tianjin, Tangshan, Shanghai and the capital, Beijing. There is also growth in provincial capitals.

A **megacity** is a city with a population of more than 10 million.

> **Exam tip**
>
> In any discussion of environmental pressures associated with rapid urbanisation provide place-specific detail and avoid restricting your answer to only one city.

As cities grow, sustainable development becomes more challenging. Demand for energy is rising. People and industries demand more water, so supplies in lakes and groundwater reservoirs fall. As land at the edge of cities is developed for new factories there is less available for farming. City sprawl separates homes from industries, increases the amount of commuting and creates traffic congestion. Domestic and industrial waste disposal facilities come under pressure, particularly from the emerging middle classes. National, provincial and city authorities are not always willing to pay for sustainable projects and services. It is reported that 90 per cent of cities have air that is harmful to breath, causing 1.6 million premature deaths each year.

Knowledge check 23

As most of the migrants are from the economically active sector, what are the implications of this age-selective migration for rural areas?

Sustainable development in China

Strategies to manage environmental problems associated with economic growth

In response to problems of deforestation, desertification and soil erosion the Chinese government has instituted one of the largest forest conservation programmes in the world, the Natural Forest Conservation Programme (NFCP). The NFCP targeted sensitive regions that had been significantly degraded over previous decades, such as upper catchments of major rivers, where extensive bans on logging are now in place. Monitoring of forest cover in China between 2000 and 2010 to evaluate the effectiveness of the programme showed that forest cover has significantly increased in 1.6 per cent of China's territory, with areas exhibiting forest gain experiencing an increase in net primary productivity (p. 11).

China's Three-North Shelterbelt Programe, involving a 4500 km green wall of trees, was initiated in 1976 to address the problem of desertification, which is particularly acute in northwest China, where annual precipitation levels are below 100 mm. The project's end date is 2050, and to date 66 billion trees have been planted. However, the non-indigenous trees planted store less carbon, use more water and are more disease prone than the indigenous species they replace.

Exam tip

It is easy to make generalisations about China, but remember that examiners are looking for detail. Learn some statistics to support your answers and note spatial variations.

Strategies to improve water, food and energy security

Water

In order to address the uneven distribution of water resources, China's South–North Water Transfer Project (Figure 16) draws water from southern rivers and supplies it to the dry north. Planned for completion in 2050, it will eventually divert 44.8 billion m^3 of water annually to the population centres of the drier north, such as Beijing.

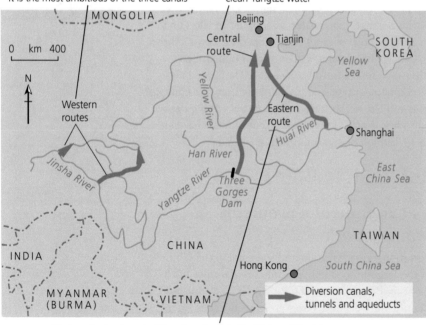

Will cut through the high Tibet plateau, linking the Mekong and Yangtze with the Yellow River. It is the most ambitious of the three canals

Will supply water for big cities like Beijing and Tianjin. Reservoir will be built to collect clean Yangtze water

Wastewater treatment will be given top priority. Water will be taken from the lower Yangtze basin where most polluting factories are

Figure 16 China's South–North Transfer Project

Another strategy designed to offer a solution for both water-scarce and waterlogged cites is the 'sponge-city initiative'. In traditional cities impermeable surfaces reduce drainage basin stores and increase flows, whereas 'sponge cities' use rain gardens, green roofs,

Exam tip

Reflect on the ways in which 'sponge city' features will influence drainage basin stores, flows and outputs using the knowledge and understanding you gained from your study of the water cycle.

constructed wetlands and permeable pavement to naturally capture, slow down and filter stormwater. That water can then replenish groundwater aquifers or be stored for future use. Thirty pilot 'sponge cities' have been selected, including Beijing and Shanghai. Xiamen, a coastal 'sponge city' in southern China, endured Typhoon Nepartak, the third most intense tropical cyclone worldwide in 2016, without waterlogging.

Food

China is the world's largest producer and consumer of agricultural products. Although distribution remains very uneven in China, all components of food security (p. 29) have been improved as a result of government policies, including:

- establishing a red line of a minimum of 120 million hectares of arable land
- setting a grain self-sufficiency target of 95 per cent, which China has maintained over the past two decades
- maintaining some of the largest grain stockpiles in the world

Energy

Clean energies made up only 13 per cent of total primary energy consumption in China in 2016, with nuclear, hydropower, wind, solar PV and biofuels accounting for 2, 8, 2, 0.5 and 0.5 per cent respectively. Hydropower currently provides the foundation of China's energy transition, but in 2017 China installed more than 52 GW of solar capacity. The Chinese leadership is keen to promote a green transition of the country's energy system because it will offer both climate change mitigation and improved energy security.

China is now the world's biggest investor in renewable energy. In 2018 China's National Development and Reform Commission (NDRC) developed a draft policy that would increase the renewable energy target from 20 to 35 per cent by 2030. The government is experiencing challenges in its attempt to slow down the construction of coal power plants (p. 41), however, the role of renewable energy is growing and simultaneously dropping in price. The newly proposed Renewable Energy Portfolio Standards for 2030 will have the power to fine companies that do not meet new pollution standards, with the proceeds used to cover renewable energy subsidies.

Strategies to improve the sustainability of urban communities

To mitigate the growing levels of air pollution the Chinese government has selected 13 cities to switch over to green vehicles. Shenzhen has an all-electric public bus fleet and plans to electrify all its taxis by 2020. Shenzhen has also launched a subsidy on private electric cars and paid for recharging stations. In less than a decade, the city reduced its air pollution levels by around 50 per cent. Shenzhen is now classified as the most sustainable city in China; it met its air quality goals in both 2016 and 2017.

China has 650 established cities, and claims to be developing 285 purpose-built eco-cities across the country. It is estimated that over 50 per cent of China's new urban developments will be labelled 'eco', 'green', 'low carbon' or 'smart'. Although there is some scepticism about levels of sustainability of these eco-cities, they will undoubtedly be an improvement on many of China's existing highly polluted, contaminated and unsustainable urban areas.

> **Exam tip**
>
> Keep up to date with changing Chinese policies, and be aware that information is managed by the government and therefore often censored.

> **Exam tip**
>
> Research the progress made by Chinese manufacturers to develop solar, wind and clean coal technology. This recent trend contradicts the conventional picture of China's poor environmental image.

> **Knowledge check 24**
>
> Name one of China's planned ecocities.

Summary

- The physical background of China ranges from mountains, high plateaus and deserts in the west, to plains, deltas and hills in the east. Climatic patterns are complex.
- China's population is the world's largest. Population distribution is uneven. The population is ageing, with a pronounced gender imbalance. Access to healthcare and education is uneven, but there have been recent improvements in both.
- China has large reserves of coal, but these are unevenly distributed and of low quality. The physical environment creates opportunities for tourism, and produces fertile soils for agriculture, but also presents climatic and tectonic hazards.
- The 'Open Door' reforms of 1978 enabled China to embrace economic globalisation. SEZs established in coastal areas stimulated economic activity. Chinese government decisions have been important in influencing the distribution and development of economic activity.
- China became the world's outsourcing capital for manufacturing although, more recently, the service sector has assumed greater importance. China plays an important role in global economic and political systems, and is the most likely rival to the USA's current dominance.
- Threats to China's environment associated with economic growth include environmental pressures, issues of water, food and energy security, and rapid urbanisation.
- The Chinese government has instituted one of the world's largest forest conservation programmes and the Three-North Shelterbelt Project to address desertification. Water security is being improved through the South–North Water Transfer Project. Food security has been improved through other government strategies. Renewables are becoming increasingly important in China's energy mix. Urban environments are being improved through promoting green transport policies and constructing eco-cities.

Economic growth and challenge: development in an African context

Definitions, measures and patterns of development

Changing definitions of development

'Development' is very difficult to define. Earlier views on 'development' emphasised economic expansion and assumed an ordered progression towards Western standards. In the 1980s and 90s, criticism of approaches to development that emphasised a growth in gross national income (GNI), industrialisation and technological advance, and the increasing awareness that development theories had not delivered success, led to a broader view of development.

A variety of concepts involving social and cultural advancement, human rights, gender equality and the role women play in development became important. The economist Amartya Sen (1999) argued that freedom is both the primary end and the

A useful working definition of **development** is 'an increase in standards of living and quality of life for an increasing proportion of the population'.

GNI per capita in US dollars is the sum of all goods and services produced in a country plus taxes and income from abroad, divided by the population. GNI was previously known as GNP.

principal means of development. Since the Brundtland Report (1987), economic, social and environmental aspects have been integrated into the broader definition of development as sustainable development (p. 31).

Exam tip

Using the correct definitions and appropriate terminology in your answers is important for strengthening AO1 marks (awarded for knowledge and understanding).

Measuring development

Simple indicators measure only one aspect of development. These include GNI, percentage of people living on $2 a day, adult literacy, life expectancy, daily calorie intake, level of malnourishment, infant mortality, energy consumption, percentage of employment in agriculture, manufacturing and services, and even the Big Mac index. None of these accurately reflects the totality of 'development'.

Composite indicators are more comprehensive because they measure more than one aspect of development by amalgamating several characteristics into one index. The human development index (HDI) is the most commonly used composite measure of development. It uses mean figures for life expectancy at birth (years), adult literacy rate (%), combined primary, secondary and tertiary gross enrolment ratio (%), and GNI per capita (listed as PPP US$). Development should increase the quality of life of people by raising the HDI to as close to 1.0 as possible. The information in Table 2 reveals that, although they vary, sub-Saharan African countries are globally marginalised in terms of their development because the HDI rankings for most sub-Saharan African countries are among the lowest in the world.

Knowledge check 25

The World Bank divides countries into four income groupings — low, lower-middle, upper-middle and high — using gross national income (GNI) per capita, in US dollars. Research the correct groupings for the African countries you are studying. How many of the countries listed as low income are in sub-Saharan Africa?

Table 2 Human development index (HDI) for selected sub-Saharan countries (2018) (1= highest level of development, 0 = lowest)

Country	Mean years of schooling	Life expectancy at birth	GNI per capita (PPP US$)	HDI	HD1 rank (out of 189 countries)
Botswana	9.3	67.6	15,534	0.717	101
Ghana	7.1	63.0	4,096	0.592	140
Kenya	6.5	67.3	2,961	0.590	142
Nigeria	6.2	53.9	5,231	0.532	157
Ethiopia	2.7	65.9	1,719	0.463	173
Democratic Republic of Congo	6.8	60.0	796	0.457	176
Niger	2.0	60.4	906	0.354	189

Other composite measures include the gender empowerment measure (GEM), which reflects gender inequality in three key areas: the extent of women's political participation and decision making, economic participation and decision making, and power exerted by women over economic resources. The human poverty index (HPI-1), which aims to capture many, but not all, dimensions of poverty in the developing world, measures variables including the probability at birth of not surviving to 40 years old, adult illiteracy, percentage of population without sustainable access to an improved water source and percentage of children under weight for their age.

Quantitative indicators are those that are measurable from collected data. Qualitative indicators are those that are not easily quantifiable. They emphasise development in terms of issues such as freedom, security and sustainability, rather than through statistics. Examples of qualitative indicators include the environmental vulnerability index (EVI), the corruption perception index (CPI) and the happy planet index (HPI).

The use of both quantitative and qualitative indicators is necessary in assessing the level of development of a given country. However, indicators can be misleading:

- Most are averages and therefore do not show how far the benefits of development are shared within a country.
- They do not show the harmful side-effects that can occur. For instance, a rise in car ownership indicates a general rise in living standards economically, but some of the population will suffer from increased noise and air pollution.
- They are sometimes incomplete, inaccurate, out-of date and, for some countries, unreliable.

Economists use the Lorenz curve and Gini coefficient to measure inequalities within countries. The closer a calculated Gini coefficient is to 1, the greater the inequality of income distribution. Countries with Gini coefficients between 0.5 and 0.7 are regarded as having unequal income distributions, whilst those having Gini coefficients between 0.2 and 0.35 are considered to have relatively equitable income distributions.

The development gap and development continuum

Although sub-Saharan Africa achieved an average reduction in its Gini coefficient (from 0.47 to 0.43) between 1991 and 2011, the region remains one of the most unequal in the world, with 10 of its 48 countries listed among the 19 most unequal in the world. Seven countries (South Africa, Botswana, Namibia, Zambia, Central African Republic, Comoros and Lesotho) are notable for the concentration of land and socio-economic assets in the hands of a few. These countries lead the region in income inequality and highlight the **development gap**. By contrast, Burkina Faso, Mali, Niger, Burundi and Guinea, characterised by communal land ownership and egalitarian access to land, rank among the most equal countries globally, and reveal more of a **development continuum**.

Exam tip

You need to appreciate that sub-Saharan Africa is made up of countries with a diverse range of physical, economic, social, cultural and political characteristics.

Knowledge check 26

Why might a sub-Saharan African country with a particular HDI rank have a different rank for its GEM?

The **development gap** refers to the widening difference in levels of development within and between countries.

The **development continuum** recognises that variations in levels of development within and between countries reveal more of a graduation than a gap.

Exam tip

Learn up-to date development statistics for at least two countries in sub-Saharan Africa to illustrate variations in levels of development both between and within countries.

Variations within countries — regional, ethnic and gender differences

Ethnic diversity and associated inequalities are a feature of many sub-Saharan African countries. Many state boundaries were drawn by colonial powers hundreds of years ago, based on raw material and water availability, with no consideration given to the different ethnic groups. See Figure 17 on page 37 of *WJEC/Eduqas A-level Geography Student Guide 5: Global Governance: Change and challenges; 21st century challenges* for an ethno-linguistic map of Africa.

Nigeria is the most populous country in Africa and one of the most ethnically diverse countries in the world. Three ethnic groups — the Hausa, Yoruba and Igbo — dominate three zones — the northwest, southwest and southeast, respectively. The ethnic gap has persisted over time in terms of access to electricity, education and wealth, with statistics for the northwest ethnic zone reporting below national mean access for these three variables. In contrast, the ethnic gap has narrowed significantly for access to more locally administered services, including improved pit latrines and potable water. Although the gender gap in education has narrowed nationally, the distribution of gains has been uneven in terms of ethnic group, with significant gender gaps persisting in the northwest.

In Kenya, the top 10 per cent richest households control more than 40 per cent of the country's income and the poorest 10 per cent control less than one per cent. Nearly every child in the Central province is enrolled in primary school, only 3 per cent of women have no education and there are about 20,000 people per doctor, whereas in the Northeastern province only one out of three children goes to school, 93 per cent of women have no education and there is one doctor for every 120,000 people.

The influence of physical factors on development

Minerals and energy sources

Sierra Leone, a small West African country and one of the poorest in the world (ranked 184th in 2017, with an HDI of 0.419), has many mineral resources including diamonds, gold, titanium, bauxite and iron ore. Mining these resources contributed about 20 per cent of GNI until the interruption of mining because of the civil war in 1991–2002. Diamond mining is Sierra Leone's second-largest employer (after agriculture), providing a livelihood for 400,000 people. Diamond mining also supplies valuable start-up capital for other economic activities and support for smallholder agriculture.

The diamonds are highly accessible near the surface of river beds, making the diamond-rich Kono district a magnet for migrant workers — mostly young, single, uneducated and unemployed men. However, due to poor governance and widespread corruption, only a fraction of this wealth returns to the areas where diamonds are mined. Many mining operations take place in remote locations outside the reach of government monitors, and close to the porous borders with Guinea and Liberia. The 'artisanal' nature of mining is unregulated, with the result that diggers are exploited by middlemen and become trapped in cycles of poverty and indebtedness.

In Sierra Leone, as in many sub-Saharan African countries, the Chinese have made significant investments, often bringing abandoned mines back into production and constructing associated infrastructure, such as railways and ports. Minerals are exported to China to meet the needs of China's industries and this guarantees an income to Sierra Leone's government. However, there are concerns that this **neo-colonial** relationship is exploitative, and although ports become centres of economic growth, they often leave inland regions severely under-developed.

Key energy sources for most sub-Saharan African countries are oil and hydropower. Oil resources in South Sudan, Mozambique and Angola are exploited with the aid of Chinese investment. Sub-Saharan African countries have tended to go for large, prestigious hydropower schemes, such as Ethiopia's 6000 MW Grand Renaissance dam (Africa's biggest) on the Blue Nile, which will provide electricity for manufacturing and a growing urban demand. Reservoirs are often not used to full capacity, and river flows are highly seasonal and increasingly suffer from irregular rainfall due to climate change.

Due to the limited economic and infrastructural resources in many sub-Saharan African countries, the extension of national grids is taking place at a very slow rate. The lack of access to sufficient and affordable energy is a major constraint on these countries' socio-economic progress. Increasingly, sub-Saharan African countries are adopting low-cost renewable technologies and bypassing fossil fuel-based industrialisation. Microgeneration is therefore increasingly meeting the energy requirements of areas far from national grids. In West Africa mini-hydropower options could soon provide up to 70 per cent of rural electricity.

Soils, relief, climate and water availability

Kenya is situated on the equator and borders the Indian Ocean. Its diverse physical geography promotes development in tourism, agriculture (Kenya's leading foreign income earner) and energy:

- Tourism — Kenya's coastline is characterised by long stretches of pristine tropical beaches bordering the Indian Ocean, offering marine-based activities and an attractive climate. 65 per cent of tourists visit the Kenyan coast. The savanna grasslands of the Great Rift Valley and the Masai Mara are among the most diversely populated wildlife areas in the world, and visited by thousands of tourists each year.
- Agriculture — Kenya's highlands have rich, fertile soil and are among the most successful agricultural regions in Africa. Kenya is the largest supplier of cut flowers to the EU, with over 50 per cent of the country's total flower production concentrated around Lake Naivasha.
- Energy — Kenya's energy sector is increasingly benefitting from its climate and geological background. It receives daily insolation of 4–6 kWh/m² and leads Africa in the number of solar power systems installed per capita. Kenya's geology also influences the location of active areas for geothermal energy — it is Africa's first and largest producer of geothermal power. These two energy sources will enable the Kenyan government to achieve its goal of universal access to electricity by 2020.

Neo-colonialism is the dominance of strong nations over weak nations, not by direct political control (as in traditional colonialism), but through economic and cultural influence.

Exam tip

Think critically about the influence of physical factors on development. They do not operate in isolation, can both promote and hinder the development process, and their influence changes over both space and time.

Knowledge check 27

Using your knowledge of the distribution of plate boundaries, explain why Kenya is an active area for geothermal energy.

The constraining effects of climate variability on development

Sub-Saharan African countries are very vulnerable to the effects of climate change because of their low adaptive capacity (p. 82). For Malawi, a land-locked country bordered by Mozambique, Zambia and Tanzania, the impacts of climate variability have an adverse effect on food and water security, water quality, energy and sustainable livelihoods:

- Floods and droughts have led to crop failures and chronic year-round food deficits.
- Crop failures result in food insecurity and malnutrition, particularly among vulnerable rural communities (where 85 per cent of Malawi's population live) dependent on **subsistence agriculture**.
- Women are impacted disproportionately because of their role as water collectors — droughts result in longer journeys to collect water.
- Floods and droughts have resulted in the disruption of hydropower generation along rivers such as the Shire River, one of Malawi's major energy sources.
- Floods result in water pollution and the increased incidence of malaria, cholera and diarrhoea.
- Drought has affected the reproduction and migration of wildlife on which the rural population depends for its subsistence.
- An increase in droughts and floods disrupts water supplies (availability, quantity and quality), which are critical for human and industrial use.
- Droughts lead to land degradation, loss of soil fertility and forest fires, which affect Malawi's forestry sector (both commercial plantations of eucalyptus and indigenous Miombo woodlands, which provide an important source of wood, fuel and food for local people).

> **Subsistence agriculture** is food production targeted at achieving self-sufficiency for a farmer's family, with little or no surplus to trade.

The influence of economic factors on development

Free trade and trade blocs

The governments of many sub-Saharan African countries often attempt to isolate themselves from global financial and trade flows, and adopt **protectionist** policies. Although the precise nature of trade relations varies from country to country, the case of Sierra Leone can be used to illustrate some of the issues that sub-Saharan African countries face, and how trade can both promote and hinder development.

In 2016 Sierra Leone's export economy was ranked 154th in the world. It has an overdependence on primary products, which are vulnerable either to low prices or to sudden shocks as global demand and prices fluctuate. Imports comprise mainly higher-value, manufactured goods and oil products. Exports in 2016, including processed crustaceans, iron ore, diamonds and titanium, amounted to US$897 million, whilst imports including rice, medicines, cars, refined petroleum and cement amounted to US$1.2 billion, resulting in a negative trade balance. The top export destinations were China and the EU, with imports primarily from China, the USA, India, UAE and the UK.

> **Protectionist** policies shield a country's domestic industries from foreign competition by taxing imports.

Sierra Leone is a member of the Economic Community of West African States (ECOWAS), which has been beneficial because tariff and non-tariff barriers have been abolished among its 15 member states. Both trade and economic activity have increased as a result, improving Sierra Leone's competitiveness in global markets. Bilateral partnerships have been negotiated with China and the UK. Although considered exploitative in their demand for iron ore, Chinese companies have financed a new hospital in Freetown, a new ministry and a rail network. Enhanced bilateral trade with the UK is expected to strengthen both economic and social development, democracy and human rights.

Sierra Leone benefits from duty-free access to their goods from the EU under an initiative known as Everything But Arms (EBA), which has been in place since 2001. EBA extends duty-free and quota-free access to imports of all products other than arms originating from the world's least developed countries. Similarly, access to the US market is governed by the African Growth and Opportunity Act (AGOA), which offers duty-free access for several products. Such benefits are offset, however, by the EU's Common Agricultural Policy (CAP). CAP subsidises EU farmers to produce food surpluses, which flood the markets of countries such as Sierra Leone and undercut the prices of domestically produced food, making it impossible for impoverished farmers to compete and make a sustainable living.

The resource curse and conflict

In theory, the export of primary commodities generates the income required for economic development to occur. Prices paid for gold, oil, diamonds and rare earths (used in many devices such as rechargeable batteries and mobile phones) are generally high. Unfortunately, the presence of these resources often creates a considerable development challenge by causing conflict — known as the 'resource curse'. Sierra Leone's diamonds (p. 49), South Sudan's oil supplies and DRC's metals have made conditions worse and not better for citizens. Human greed over DRC's natural resources (coltan and other minerals, including diamonds) has attracted militia groups from other countries, such as Uganda's Lord's Resistance Army. Between the mid-1990s and 2010, millions of 'conflict refugees' fled across the DRC's unmarked borders into neighbouring countries, such as the Central African Republic.

Exam tip

The resource curse is an example of a political driver of international migration, a synoptic topic that you will have covered when studying processes and patterns of global migration, part of the Global governance theme.

Influence of MNCs

FDI (p. 26) can both promote and hinder a country's development. Sierra Leone's substantial mining wealth makes Sierra Leone attractive for FDI. Although FDI was severely impacted by both the civil war (1991–2002) and the Ebola outbreak in 2014, it has increased to reach US$560 million in 2017. The shortage of skilled labour, the lack of infrastructure, the slow legal system, the high level of corruption, political violence and serious social disorder due to socio-economic disparities are, however, major obstacles to FDI.

The nature of Chinese investment in Sierra Leone, although important in promoting infrastructural and social improvements (p. 52), does not appear to be leading to significant employment generation and effective poverty reduction in the long term. Management roles are reserved for Chinese personnel and there have been allegations of child labour exploitation.

Influence of tourism

An increasing number of sub-Saharan African countries look to tourism as a catalyst for economic development. Kenya's tourist industry illustrates both the positive and negative impacts of tourism on development (Table 3).

Table 3 The positive and negative impacts of Kenya's tourist industry

Positive impacts	Negative impacts
■ Kenya earned $1.2 billion in 2017 from tourism. ■ Employment has risen. 1.1 million direct, indirect and induced jobs were supported by the industry in 2016, or 9.2 per cent of the country's total employment. ■ Tourism brings money in to the local economy (**multiplier effect**). Local farmers benefit from supplying hotels with food. ■ Income pays for conservation projects or the running of national parks, which protect plants, animals and the natural landscape. ■ Money from tourism goes towards projects for local facilities, such as schools, health clinics and improved roads. ■ Tourism has kept many young people in the area, reducing the dependency ratio (knowledge check 7, p. 100).	■ Most of the income from tourism is kept by the big travel companies (**leakage**). Less than 2 per cent of the money spent at the world-famous Masai Mara National Reserve benefits local Masai people. ■ Jobs for local people are poorly paid and seasonal. ■ Tourism is negatively impacted by shocks, such as the 2008 global recession, the political instability associated with prolonged electioneering in 2002 and 2013, and the threat from extremists linked to the militant group Al Shabaab. ■ Tourism creates conflicts between tourists and locals. In Mombasa, hotels have forbidden public access to the beaches. This has stopped the locals using the beaches.

The **multiplier effect** occurs when an increase in spending produces an increase in national income and consumption that is greater than the initial amount spent.

Leakage is revenue generated by tourism that is lost to other countries' economies.

Influence of fair trade

One initiative to improve the terms of trade for sub-Saharan African countries is the Fairtrade movement, which seeks to obtain a fair price for a wide variety of exported commodities.

The Oserian plantation in Kenya is Africa's largest producer of roses and carnations (p. 55). From a small vegetable plot with only six employees, Oserian evolved into the first cut-flower farm in the country. It is now a flagship Fairtrade project producing a million stems a day for some of Britain's biggest supermarkets. Fairtrade earnings are invested into a communal fund for workers, farmers and the community to improve their social, economic and environmental conditions. However, allegations of corruption and misappropriation of funds have been levelled at Oserian. Criticisms include employees working in stifling greenhouses and being paid well below Fairtrade's 'living wage'. In its defence, Oserian argues that most employees live in free accommodation, with benefits including electricity, healthcare, schooling and transport, bringing the value of a permanent worker's pay to above the living wage.

The influence of political, social and cultural factors on development

Political factors

Strong, accountable and transparent governance should promote all aspects of economic, social and environmental development. Unfortunately, in many sub-Saharan African countries, corruption is institutionalised and widely prevalent. The decade-long civil war in Sierra Leone (p. 49) further deepened patterns of bribery and caused development failure by affecting the confidence of overseas investors. Sierra Leone was ranked 130th in **Transparency International's corruption perception index** in 2017. To improve transparency levels in diamond mining, Sierra Leone, with the support of the UK, was a founder member of the Kimberley Process, a multi-party initiative that has developed a system of rough-diamond certification aimed at assuring buyers that they are purchasing legitimate diamonds.

Colonial links with sub-Saharan African countries are often seen as exploitative. For example, when the Democratic Republic of Congo gained independence from Belgium, much of the country's vast mineral wealth had been 'looted' and, in terms of human capital, only 16 of its citizens held degrees. However, Sierra Leone's ex-colonial donor, Britain, is helping Sierra Leone build its fragile democratic institutions and is promoting good governance. China's neocolonial infrastructural influence is quite different (p. 50).

In terms of global organisations, the UN has been a major driver of social development in all sub-Saharan African countries through its **MDGs**, now superseded by the **SDGs**.

Social factors

Development aid has enabled countries to make good progress in education and health. Universal primary education is technically available to all children, although sub-Saharan Africa still has the highest proportions of children out of school — 21 per cent of primary age, 34 per cent of lower secondary age and 58 per cent of upper secondary age. There have been improvements in female enrolment, particularly under the MDGs and SDGs, but girls still suffer severe disadvantages and exclusion, particularly in rural areas and among the rural poor. Female education enables women to move into the labour market and increase the productive capacity of the labour force.

The transformational effect of female empowerment on development can be seen in Rwanda. A new constitution was passed in 2003 decreeing that 30 per cent of parliamentary seats must be reserved for women. By 2018 Rwanda led the world in female representation, with 64 per cent of seats held by women. A compulsory education programme ensures gender equality in both primary and secondary schools. Women can own and inherit property and are active leaders in the business community. However, although national mandates are reducing violence against women, Rwanda remains fundamentally a patriarchal society, with high levels of domestic violence.

Transparency International's corruption perception index ranks 180 countries by their levels of public-sector corruption, as perceived by businesspeople and other experts.

MDGs stands for the UN's millennium development goals, and **SDGs** for the UN's sustainable development goals.

Knowledge check 28

Outline the UN's MDGs and SDGs.

The impact of development on the environment

Consumerism and the impact of the exploitation of natural resources

Economic development and associated poverty reduction in sub-Saharan African countries has led to an increase in the number of middle-class people. Increased income means more spending power and aspirations for more Western-style consumption of white goods, fast food, petrol and cars, all of which contribute to increased waste and emissions.

In Kenya, rapid urbanisation has generated increasing amounts of wastewater and refuse, exceeding the capacity of cities' waste collection and treatment systems. The growth in car ownership has caused increased congestion and emissions in cities such as Nairobi. However, some countries are making attempts to address these negative impacts. Rwanda has adopted a green growth strategy, with initiatives such as a ban on plastic bags, maintenance of minimum forest cover, promoting the use of green energy systems and reducing the use of bottled water.

Mining operations are mostly associated with negative environmental impacts. Monitoring of environmental changes in southwestern Sierra Leone resulting from the mining of titanium dioxide, using Landsat images and field hydrological and biophysical data, has shown that reservoir construction for mining has led to flooding, deforestation and the creation of waste tips. The negative environmental impacts associated with oil exploitation in the Niger Delta and gold mining in Ghana have also been widely documented.

Impacts of agro-industrialisation

Most sub-Saharan African countries have seen a shift from traditional food systems and small-scale farming towards agribusiness. One example is the Kenyan cut-flower industry (p. 53), which has been accused by NGOs and environmental agencies of having a negative impact on the country's environment through:

- the unsustainable use of water resources, with concerns about the reduction in the water level of Lake Naivasha, and its possible disappearance over the next few decades
- the use of agrochemicals, which destroy soil structure
- contamination of water and soil by harmful chemicals and pesticides
- destruction of wetlands and original natural habitats

Some more positive aspects of agro-industrialisation include the following:

- Many farms are researching organic methods of pest management.
- The management of water resources is improving through the recycling of wastewater and collection of rainwater, tree replanting and/or forest preservation. For example, the Lake Naivasha Riparian Association, a voluntary, self-regulating body, now monitors the amount of water extracted, and water resources have improved.
- In Ghana the MNC Cadbury is safeguarding supplies of high-quality cocoa beans to promote higher yields and the sustainability of cocoa cultivation.

> **Exam tip**
>
> Make the link between unsustainable agricultural practices and desertification.

Manufacturing industries

The impacts of manufacturing in sub-Saharan African countries are invariably negative, although attempts are being made to address this. Industrial manufacturing is relatively strong in Kenya, accounting for 20 per cent of economic activity and

12.5 per cent of all formal jobs in the economy. Notable sectors include chemicals, metals, pharmaceuticals, furniture, leather goods and motor vehicles. Foreign companies often set up their businesses with disregard for environmental pollution. Industrial waste is responsible for river pollution and illness within the local communities. Manufacturers of electrical and electronic equipment generate increasing amounts of e-waste.

Challenges of desertification

Causes of desertification

The causes of desertification (p. 28) can be seen in many countries of the **Sahel**.

The primary causes of desertification are natural:

- Climate change is associated with precipitation becoming less reliable, seasonally and annually. Average annual departures from normal rainfall increase progressively from south to north in the Sahel (Figure 17). Drought periods can extend to several years.
- Vegetation is increasingly water-stressed and dying off, exposing bare soil.
- Bare soil is vulnerable to erosion by wind and water.
- Short, intense precipitation results in rapid runoff (infiltration-excess overland flow), reducing the soil moisture store.

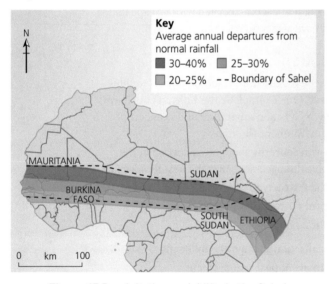

Figure 17 Precipitation variability in the Sahel

Human factors reinforce the physical causes, leading to positive feedback. Desertification intensifies and spreads further due to:

- overcultivation and an increase in large-scale, mechanised agriculture, with accompanying soil compaction and increased runoff
- overgrazing by goats, sheep and cattle, which destroys vegetation
- deforestation and the use of trees for fuel and construction, resulting in fewer roots to bind the soil and reduced interception
- population growth and pressure, forcing rural dwellers onto more marginal land

The **Sahel** is a semi-arid region extending across north Africa from Senegal eastward to Sudan. It forms a transitional zone between the arid Sahara Desert to the north and the belt of humid savanna to the south.

Exam tip

Consider the different specialised concepts you could apply to an exam question on desertification. Where could you integrate the specialised concepts of equilibrium, feedback, sustainability and threshold into your discussion?

Albedo is the proportion of incoming solar radiation that is reflected by a surface.

Knowledge check 29

What changes in **albedo** occur when vegetation cover is replaced with bare ground? What feedback mechanisms may be triggered as a result?

Knowledge check 30

Distinguish between infiltration-excess and saturation-excess overland flow, which you would have studied when covering causes of excess runoff in your study of the water cycle.

- the change from nomadic to sedentary agriculture — a sedentary population is socially more developed, with improved access to healthcare and education, but areas are used continuously and easily exceed their **carrying capacity**
- civil wars, for example in South Sudan, which can intensify the problems

Carrying capacity is the maximum population size that the environment can sustain.

Consequences of desertification

The consequences of desertification are complex. Figure 18 summarises the impacts of desertification on ecosystems, populations and landscapes. Population pressure and conflict over limited resources has occurred in northeast Kenya due to pastoralists migrating from Mandera to neighbouring Wajir county in search of water and pasture.

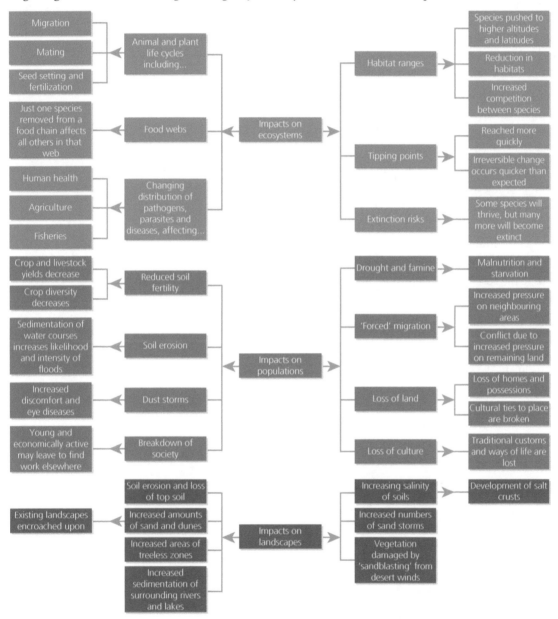

Figure 18 Impacts of desertification on ecosystems, landscapes and populations

Strategies to address the causes and consequences

Strategies use a range of technologies and are implemented at a local, national and international level.

The Australian NGO World Vision Australia (WVA) has been supporting the Dryland Development Programme (DryDev) in Ethiopia and Kenya. DryDev assists dryland farmers in the transition from subsistence and food aid to sustainable rural development. Since 2013 funding from the Dutch government, WVA and partners, REST (Relief Society of Tigray) and EOC (Ethiopian Orthodox Church) has provided 40,000 farmers in Ethiopia with solutions to address land degradation. In one valley in Tigray, well-known for its dry conditions and steep slopes, the community have been supported in building stone terraces, weirs, small dams, contour bays and soakage pits, planting trees, minimising animal browsing and allowing trees to recover and protect the soil. Since DryDev began, yields have doubled, the irrigation area has expanded from 10 to 30 hectares, and more intensive cropping can be practised, bringing more secure and better food prices.

At the national level, Kenya's government has introduced policies such as:
- subsidies on farm inputs, especially fertilisers
- purchase of maize from farmers at prices above market prices to provide incentives to producers
- providing subsidies to maize millers, to bring down the prices of maize to consumers
- agricultural replanting schemes
- diversification, for example into beekeeping
- developing opportunities for small business enterprises for women
- raising the levels of strategic food reserves, to stabilise maize prices
- providing a fund for the purchase of livestock from drought-stricken areas
- contributions to the costs of social amenities, for example free education and reduced costs of health facilities, to provide more disposable income to spend on food

In 2007, an African-led initiative was launched with the ambition to grow an 8000 km 'Great Green Wall' across the entire width of Africa, from Senegal in the west to Djibouti in the east. A decade in, and roughly 15 per cent underway, the initiative is already bringing life back to Africa's degraded landscapes at an unprecedented scale, providing food security, jobs and a reason to stay for the millions who live along its path. The Great Green Wall aims to address the combined threats of climate change, drought, famine, conflict and migration.

Strategies to promote development

National governments

Every sub-Saharan African country has a 5-year development plan that covers health, education, water, trade and infrastructure, and identifies its priorities. Aid agencies are required to work within this plan. Critics suggest that there is pressure on governments to support particular policies for development in order to ensure they get funding from donors such as the World Bank. Some donors, including DFID, will only support particular neo-liberal policies, and that in turn influences the approach taken by a country's government.

Through its Growth and Transformation Plan (GTP), Ethiopia aims to reach lower middle-income status by 2020–23. During the first 5 years (GTP I 2010/11–2014/15),

> **Exam tip**
>
> Note that strategies operating at different scales are often integrated. Global agencies work with local partners and liaise closely with national governments.

Ethiopia's economy registered an annual average GNI growth rate of over 10 per cent, which is well above the required annual average growth rate of 7 per cent set for low-income countries to achieve sustained and inclusive growth, realise UN SDG targets and eradicate extreme poverty.

International aid agencies, NGOs and microfinance schemes

Aid contributes to national development plans by providing infrastructure, improving governance and accountability, encouraging capacity building, providing basic services and technological support, encouraging trade and increasing empowerment. There is increasing interest in the use of 'smart aid' — programmes that focus on bottom-up projects in order to increase the effectiveness of each £ or $ given in aid. Examples include education (particularly of girls), safe water and sanitation projects, the provision of small loans (usually of less than US$200) to poor people, the eradication of diseases and research to discover new crop varieties resistant to virus and drought.

In most sub-Saharan African countries, women are often at a disadvantage because the structural, financial institutions of the country do not allow women to develop as entrepreneurs. Microfinance schemes offered by NGOs, such as BRAC working in Sierra Leone, provide microloans and enterprise loans to women (and some men) to help them engage in income-generating activities.

The World Bank and IMF

International organisations such as the World Bank and International Monetary Fund (IMF) aim to encourage national governments to reject protectionist policies and adopt a free-trade approach. The World Bank provides direct grants to sub-Saharan African countries, but any support is governed by strict conditions. The IMF channels loans from wealthy countries to those that apply for help, but in return recipients must agree to operate free market economies open to outside investment. Often their role is controversial because they impose strict financial conditions on borrowing governments, which may be required to cut back on social programmes such as healthcare, education and sanitation.

Exam tip

When selecting examples of aid, choose programmes that have been rolled out more widely and provide evidence of economic growth and development, rather than those that offer only limited and small-scale benefits. Be aware that some aid programmes have led to aid dependency.

Summary

- 'Development' can be defined and measured in a variety of ways, and there are variations in development both between and within sub-Saharan African countries.
- Development is influenced by a complex interplay of physical, economic, political, social and cultural factors.
- Resource exploitation (e.g. mining), consumerism and agro-industrialisation have often had a negative impact on the environments of sub-Saharan African countries. Increasingly, 'green' strategies are being introduced to address these impacts and achieve more sustainable development.
- Desertification is a significant problem in areas such as the Sahel. Strategies use a range of technologies and are implemented at the local, national and international level.
- Development is promoted at different scales by national governments, international aid agencies, NGOs, microfinance schemes, the World Bank and IMF.

■ Energy challenges and dilemmas

The classification and distribution of energy resources

Classification of energy resources

Energy resources are classified as either non-renewable or renewable.

■ The exploitation of non-renewable resources leads to their exhaustion because their rate of formation is very slow. Non-renewable resources include hydrocarbons (the fossil fuels) and uranium ore, which is required to generate nuclear power.

■ Renewable energy resources can be consumed in any given time period, provided current use does not exceed net renewal rates during the same period. Renewable resources include wind, solar power, hydropower, geothermal energy, wave and tidal power, and biomass. Renewable resources can be subdivided into critical or recyclable resources, including biomass energy (which requires management to ensure sustainable use), or non-critical or everlasting resources such as tides, waves, wind, running water and sunshine.

A reserve is the proportion of a resource that can be exploited under current economic conditions, and with available technology. Recoverable (proved) reserves refer to the amount of an energy resource likely to be extracted for commercial use that is economically and technologically viable for extraction. Speculative reserves or stocks are deposits that are currently not economically viable, or have not yet been explored.

Primary energy resources are raw materials that are used in their natural form to produce power, such as coal, wood and sunlight. Secondary energy resources involve converting a primary energy source into a new form, such as coal into electricity.

> **Exam tip**
>
> Using the correct definitions and appropriate terminology in your answers is important for strengthening AO1 marks (awarded for knowledge and understanding).

Global distribution of fossil fuel stocks and reserves

Fossil fuel stocks and reserves are distributed unevenly. Deposits of oil, gas and coal are a coincidence of geological history and international boundaries. Reserves run down over time, as is the case with gas from the UK's once abundant North Sea oil and gas supplies. Remaining oil and gas will be increasingly concentrated in the Middle East, North America and the Russian Federation over the next 30 years.

Coal

Coal is the most widely distributed and abundant fossil fuel in the world. By region, Asia Pacific holds the most proved reserves, split mainly between Australia, China and India. The USA remains the largest single reserve holder (Table 4).

Table 4 Coal production in 2017 and proved reserves at the end of 2017

Country	Production (million tonnes)	Total proved reserves at the end of 2017 (million tonnes)	R/P ratio in years
China	3523.2	138,819	39
India	716.0	97,728	136
USA	702.3	250,916	357
Australia	481.3	144,818	301
Indonesia	461.0	22,598	49

Source: **BP** Statistical Review of Energy 2018

Natural gas

Natural gas provides a source of relatively clean and cheap energy. The largest reserves are concentrated in the Russian Federation, and in Iran and Qatar in the Middle East (Table 5). The USA has been the world's biggest producer of natural gas since 2010.

Table 5 Natural gas production in 2017 (billion m^3) and proved reserves at the end of 2017 (trillion m^3)

Country	Production (billion m^3)	Proved reserves at end of 2017 (trillion m^3)
USA	734.5	8.7
Russian Federation	635.6	35.0
Iran	223.9	33.2
Canada	176.3	1.9
Qatar	175.7	24.9

Source: **BP** Statistical Review of Energy 2018

Oil

Although the USA is now the world's biggest oil producer, the greatest concentration of recoverable reserves are recorded in Venezuela (47.3 thousand million tonnes), Saudi Arabia, Canada and Iran (Table 6).

Table 6 Oil production in 2017 and proved reserves at end of 2017

Country	Production (thousands of barrels daily)	Proved reserves at end of 2017 (thousand million tonnes)	R/P ratio in years
USA	13,057	6.0	10.5
Saudi Arabia	11,951	36.6	61.0
Russian Federation	11,257	14.5	25.8
Iran	4,982	21.6	86.5
Canada	4,831	27.2	95.8

Source: BP Statistical Review of Energy 2018

R/P ratio (reserves to production ratio) refers to the reserves remaining at the end of a given year divided by the production in that year. The ratio indicates the length of time that remaining reserves would last if production were to continue at that level.

Alternatives to conventional fossil fuel sources

Recent increases in US oil and gas production are due to new drilling techniques, such as horizontal drilling and **hydraulic fracturing (fracking)**, which have unlocked large quantities of oil and gas from shale rock, especially in Texas and North Dakota (Figure 19(a)).

Another unconventional fossil fuel source is tar sands (oil sands), a combination of clay, sand, water and bitumen (a heavy, black, viscous oil). Tar sands can be mined and processed to extract the oil-rich bitumen, which is then refined into oil. Much of the world's oil (more than 2 trillion barrels) is in the form of tar sands, although it is not all recoverable. The largest deposits are found in Canada (Alberta) (Figure 19(b)) and Venezuela.

Deepwater oil, found well offshore and at considerable oceanic depths, is also classified as an unconventional fossil fuel because of the development of complex technology to access it. Brazil's deepwater oil came onstream in 2009 from oil and gas fields located more than 200 km offshore. Rigs drill more than 2000 m below the sea surface and many more thousands of metres below the seabed (Figure 19(c)).

Hydraulic fracturing (fracking) is a technique designed to recover gas from shale rock reserves found 1000–4000 m below the ground surface. It involves drilling a borehole into the Earth and then injecting a high-pressure mixture of water and chemicals to fracture the rock, which releases the gas and allows it to flow into the borehole.

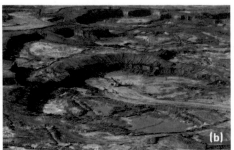

Figure 19 Three unconventional fossil fuels: (a) US shale gas, (b) Canadian tar sands, (c) Brazilian deepwater oil

Exam tip

The risks of drilling for deepwater oil can be considered in synoptic terms. There are important links to the sovereignty of ocean resources, managing oceanic pollution *(Global governance)*, and risks from extreme weather events in the rough seas far off the Brazilian coast and from hurricanes in the Gulf of Mexico *(Weather and climate)*.

Renewable alternatives to conventional fossil fuel sources include nuclear, solar, ocean, wind, biomass, hydrological and geothermal energy (p. 60).

Physical factors determining the supply of energy

Geological factors

Fossil fuels are formed over geological time from the decayed remains of animals and plants. They are concentrated in specific locations where geological conditions have created oil and gas traps, and the formation of deltaic swamps allowed coal to be formed. Geological factors also influence the location of active areas for geothermal energy — for example, 87 per cent of Iceland's demand for hot water and heat comes from geothermal energy.

Geology can be an important indirect factor — for example, large nuclear power stations are best located on geologically and seismically stable foundations.

Climatic factors

Certain forms of renewable energy are constrained by climatic factors. Solar power requires high insolation rates; wind power relies on high, constant wind speeds characteristic of areas affected by westerly wind belts; and hydropower is usually linked to areas of high precipitation.

Relief factors

Relief is an important consideration for the generation of hydropower. The deep, narrow valleys of the west slopes of the Sierra Nevada in California provide sites for dams and reservoirs, for example at Shasta in the Upper Sacramento River Basin. Relief is also important for providing a 'head' of water, which is stored and then released to drive turbines and generate hydropower.

Locations with favourable conditions for sustainable energy generation

Certain locations provide favourable conditions for sustainable energy generation from waves, tides (tidal power is restricted to a few estuaries with a very large tidal range, such as the River Severn) and biofuels.

> **Exam tip**
>
> AO2 marks are also earned by making synoptic links with other parts of the A-level specification, such as the water cycle. Over time, changes in the volume of glacier ice associated with climate change will lead to large changes in the hydrology of glacial rivers in countries such as Iceland, with important implications for the hydropower industry.

> **Exam tip**
>
> It is important to apply your knowledge and understanding of physical factors influencing the supply of energy in order to earn AO2 marks. For example, consider how these factors vary over space and time.

The changing demand for energy

Changing global patterns of energy demand

During the twentieth century, the global demand for energy increased tenfold. Despite a brief downturn associated with the global recession of 2008–09, future trends indicate a sustained rise in **energy demand**. By 2020 **energy consumption** (Figure 20) is expected to be approximately double the level in 1980.

Demand will be driven by a combination of population growth, economic development and increased living standards, particularly in the **NICs**, **BRICs**, **MINTs** and Middle Eastern states. However, energy demand within developed economies such as Europe, North America and Japan is predicted to stabilise, or in some cases decrease (see the example of Sweden on p. 70). By the 2030s, India is predicted to emerge as the world's largest growth market for energy (p. 30), with Africa playing an increasingly important role in driving energy demand.

> **Exam tip**
>
> When asked about the changing demand for energy, remember that demand may decrease as well as increase. Technological factors are increasingly associated with a reduction in energy demand, due to greater efficiency.

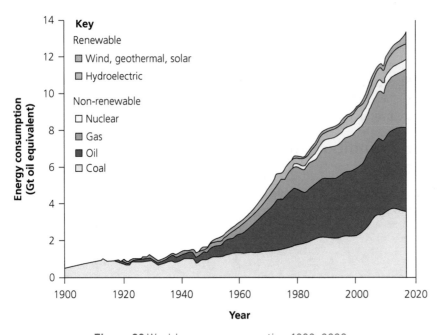

Figure 20 World energy consumption 1900–2020

Energy demand is the need or desire for energy.

Energy consumption is the availability and use of energy

NIC is the acronym for a newly industrialised country.

BRICS is the acronym for five major emerging economies: Brazil, Russia, India, China and South Africa.

MINT is the acronym for the emerging economies of Mexico, Indonesia, Nigeria and Turkey.

> **Exam tip**
>
> Using the correct definitions and appropriate terminology in your answers is important for strengthening AO1 marks (awarded for knowledge and understanding).

Economic factors influencing the demand for energy

There is a strong positive correlation between GNI per capita (p. 46) and energy usage. This is particularly the case for the economies of NICs, BRICS and MINT, where energy for manufacturing is an important driver of growth. Economies such as China's have, until recently, experienced economic growth of about 10 per cent per year. Industries such as steel, chemicals and plastics depend on energy for both heating and cooling. Growing international trade has led to the transport of goods by air, sea and land.

Demographic and social factors influencing the demand for energy

As people acquire more wealth and the middle class expands in countries such as India and China, there is a growing demand for appliances for cooking, air conditioning and lighting, which use electricity. In emerging economies car ownership becomes an aspiration and is rising rapidly in countries such as China. If China's vehicle ownership rate reaches the US level of 800 vehicles per 1000 people, then its total vehicle population would approach one billion — more than four times the present number of vehicles in the USA.

Global population continues to grow, with a higher proportion of people living in cities, which use more energy than rural areas. Worldwide power consumption for air conditioning alone is forecast to surge 33-fold by 2100. The USA uses as much electricity to keep buildings cool as the whole of Africa uses for everything. Sixty per cent of growth in expected energy consumption is directly related to **urban sprawl**.

Urban sprawl is the spread of population away from central urban areas into low-density, often car-dependent communities, in a process called suburbanisation.

Technological factors influencing the demand for energy

For many of the economic and social reasons above, technology has produced equipment that requires energy. The most rapidly growing area of demand is for electricity to support the massive servers belonging to organisations such as Microsoft and other 'cloud' technology. Servers need electricity to run and to be kept cool (Microsoft recently located one of its data centres in the North Sea near the Orkney islands). The number of electronic devices owned in the world increases daily.

> **Exam tip**
>
> In order to earn AO2 marks consider how factors influencing the changing demand for energy are interrelated. They will vary spatially, temporally and according to scale.

The global management of oil and gas

Managing supply and demand through transfers, storage and pricing

With both oil and natural gas resources being fossil fuels, a major challenge is the management of global oil depletion. Easy-to-produce oil is running out. Oil production has declined in areas such as the North Sea, and fewer giant oilfields are being discovered. However, technological developments have enabled oil and gas exploration to take place in more extreme environments — ocean depths (p. 62), areas with extreme weather conditions (e.g. the Gulf of Mexico where the Deepwater Horizon disaster occurred), fragile environments (e.g. Alaska and Greenland) — making their management more challenging.

Globally, areas of oil and gas production do not correlate with areas of consumption. Energy pathways have been created to allow transfers between producers and consumers. Transfers of oil are considerable (Figure 21). Gas is transferred either directly through pipelines (e.g. those used to export Russian gas to Europe) or converted into a liquid form (LNG) and moved by tankers (e.g. from the Middle East, Trinidad, Nigeria and Indonesia). There are plans for China to build a pipeline to bring natural gas from Iran to Pakistan. This project is politically contentious due to terrorism and instability in Pakistan and strained US–Iran relations.

Knowledge check 31

What is LNG and what issues are associated with it?

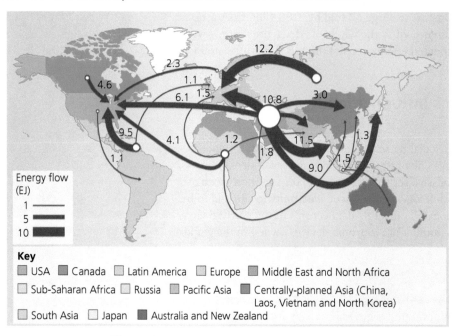

Figure 21 Global trade in oil and oil products

Exam tip

The study of energy supply and demand allows geographers to think synoptically and earn AO2 credit in exam answers. Physical geography has determined the location of oil and gas reserves, whilst human geography has influenced the locations where these resources are needed.

The Global Oil and Gas Storage Market regulates the storage of crude oil and natural gas. One of the major drivers for storage was the recent (2015–2017) drop in oil prices, because a reduction in supply of oil increases its price. Oil is also stored for strategic reasons, usually in underground salt caverns. Global storage capacity of about 7–8 billion barrels of oil is concentrated in global trading hubs such as Singapore,

New York and Rotterdam. The USA has the largest strategic oil reserves in the world. Natural gas storage underground is more complex, but is used to meet excessive demand such as during extreme winter weather conditions.

Storage is linked to the pricing of oil and gas. Oil exploration is a long-term investment, and the recent fluctuations in oil prices (US$150 a barrel in 2009 and US$50 in 2018) have been difficult for oil companies to manage. Below a given oil price, exploration and production become uneconomic.

Management by MNCs and national governments

Oil and gas industries are divided into three components: upstream (exploration and production), midstream (storing, marketing and transporting) and downstream (refining, distribution and retailing). Major MNCs (Table 7) are important players in the oil and gas industries, and have up-, mid- and downstream operations. They have a high degree of influence in terms of expertise and have a strong presence in many oil-producing countries, although they need to operate within their host country's rules.

Many companies are state-owned (and, strictly speaking, are not MNCS). Most state-owned companies spearhead upstream exploration and production. As oil has been discovered in non-OPEC countries, such as within South America and Africa, government influences have increased. Some, such as Russia, totally control foreign MNCs; others, such as Equatorial Guinea's state-owned oil company GEPetrol, are corrupt and have a negative influence on oil development.

Table 7 The world's largest oil and gas companies based on 2017 revenues

Company	2017 revenue (US$ billions)
China Petroleum & Chemical Corporation (Sinopec)	362
Royal Dutch Shell plc	305
China National Petroleum Corporation	269 (2016 figure)
BP plc	240
ExxonMobil	237
Vitol Holding BV	181
Total SA	149
Chevron Corporation	134
Gazprom	113
Rosneft Oil Co	104

Source: hydrocarbons-technology.com (2018)

Management of oil supplies by OPEC and national governments

The Organization of the Petroleum Exporting Countries (OPEC) is responsible for 40 per cent of the world's oil production and owns over 80 per cent of the world's proven oil reserves, the bulk of which are in the Middle East. The organisation was founded in Baghdad in 1960. There are currently 14 member countries, including Saudi Arabia, Iran, Iraq, Libya, Venezuela and Nigeria. As a cartel, the aim of OPEC is to coordinate and unify the policies of its member countries and ensure the stabilisation of oil markets in order to secure an efficient, economic and regular supply to consumers, a steady income to producers and a fair return on capital for those investing in the oil industry.

Many countries outside OPEC are major oil producers, notably the USA, Russia and China. Some governments, such as Norway, exercise sound stewardship and investment strategies; others lack competence and expertise.

Knowledge check 32

Identify the main players in the global energy market.

Problems associated with extraction, transport and use of energy

Environmental problems

There are many problems associated with the extraction of fossil fuels. Underground coal mines lead to surface subsidence and produce toxic waste and water. Opencast pits scar the landscape and, although legislation may require restoration of sites, the new ecosystems are of low quality. Oil infrastructure from large oilfields visually pollutes a large area. Oil spills at production sites (p. 66) are ecologically disastrous. Natural gas is generally seen as the cleanest of fossil fuels in greenhouse gas terms, but flare-off as a waste product of oilfields causes major environmental problems. Unconventional sources of oil and gas such as tar sands and shale gas may lead to water contamination, the threat of earthquakes and environmental degradation. Biofuel production is associated with an increased risk of deforestation.

The transport of energy, particularly along pipelines and tanker routes, has resulted in oil spills, with disastrous consequences for the environment. One well-known example was the Exxon Valdez oil spill in Prince William Sound, Alaska in 1989. Three decades later, oil can still be found on beaches, and the herring fishing industry has yet to recover.

The burning of fossil fuels to generate energy has led to an increase in carbon output, contributing to climate change. Coal is the most polluting source of energy, with the burning of coal leading to acid rain and smog. Some renewable forms of energy are associated with environmental problems: wind farms are considered unsightly by many people and contribute to bird deaths; and hydropower generation often involves drowning vegetation, which produces methane, a far more active greenhouse gas than CO_2. Nuclear power generation is associated with environmental risks, as demonstrated by the Chernobyl (1986) and Fukushima (2011) nuclear disasters.

Political problems

There are many political problems associated with the extraction and transport of fossil fuels, particularly oil and gas, with unstable suppliers (Libya, Iraq, Iran) and volatile pathways (Russian gas to the EU is delivered by four pipelines, three of which cross Ukraine). Political instability in areas where oil and gas resources are concentrated, such as Libya and Crimea, creates challenges for managing the transfer, storage and pricing of oil and gas. Public protests, such as anti-fracking campaigns, have delayed plans for the extraction of shale gas in the UK, with planning permission granted only recently.

Exam tip

To earn AO2 marks consider how the problems associated with energy are interrelated. As crude oil is the world's most actively traded commodity in the financial markets, it is highly susceptible to economic shocks, such as that caused by the loss of oil production from BP's Deepwater Horizon spill in 2010. This event also caused a costly environmental disaster.

Technological problems

The more electrical energy supplied by renewables, the more unstable national grids become because some sources, such as wind and solar, only produce electricity intermittently. As more renewables come online it becomes more difficult to manage fluctuations in the grid. The technology for large-scale energy generation from the ocean is unproven. Only the most developed countries have the research facilities and funding to develop new technologies. Carbon-capture technology for removing CO_2 from the atmosphere is unproven and complex (p. 72).

In terms of low technology, fuelwood is still a major source of heating and cooking energy in developing countries. Wood burning in confined domestic spaces is one of the greatest sources of ill health.

Economic problems

Fossil fuels are ultimately finite, and most world regions are heavily dependent on them.

Radical changes to the renewables industry are required to compete on the scale necessary to displace coal, oil and natural gas. Alternative forms of energy need massive investment in research and development, and almost all forms are expensive to build. Construction costs for the new generation nuclear plant at Hinkley Point in Somerset are estimated to be £20 billion.

Much of the oceans' energy potential is yet to be developed. Wave and tidal energy are at the developmental stage, so costs are high. Oceanic sources have low energy densities and so large numbers of devices are needed to harness this energy, making it expensive.

Hydropower is one of the most established alternative sources of energy. Although it generates a cheap form of electricity compared with other renewables, the initial cost of infrastructure can be high. The Three Gorges Dam in China, for example, required US$25 billion.

Energy mixes and development

Local scale — sustainable energy microgeneration

Due to the limited economic and infrastructural resources in many developing countries, the extension of national grids is taking place at a very slow rate. Microgeneration is therefore increasingly meeting the energy requirements of areas far from national grids. It is estimated by the World Bank that 89 million people in Africa and Asia already own at least one solar-powered product, and about 15 million off-grid households are predicted to have solar-powered TVs by 2020. In the drought-stricken Kenyan region of Turkana, Practical Action has installed solar panels that power water-drawing pumps for up to 12 hours a day.

National scale — the energy mix of different countries

As a country develops and energy demand increases, its energy mix will change. This can be demonstrated using a model of energy transition (Figure 22). In rural areas of LICs energy consumption is low and based on burning fuelwood and other biomass. As economies develop, increasing manufacturing (as in China following its Open-Door Policy), rapid urbanisation and rising living standards lead to growing energy

Exam tip

If one of your options is *Development in an African context*, consider the use of energy microgeneration to access water in order to address the causes and consequences of desertification.

Microgeneration refers to small-scale systems that generate electricity and/or heat for domestic dwellings and small businesses.

Energy mix refers to the combination of energy sources used to meet a country's energy demand. The exact mix varies from country to country according to energy availability, the security of supplies, national and international legislation, and sociocultural preferences.

demand and a broadening of the energy mix. Advanced economies dominated by tertiary and quaternary activities increasingly depend on secondary energy supplies, such as electricity, generated from a wide mix of fuels.

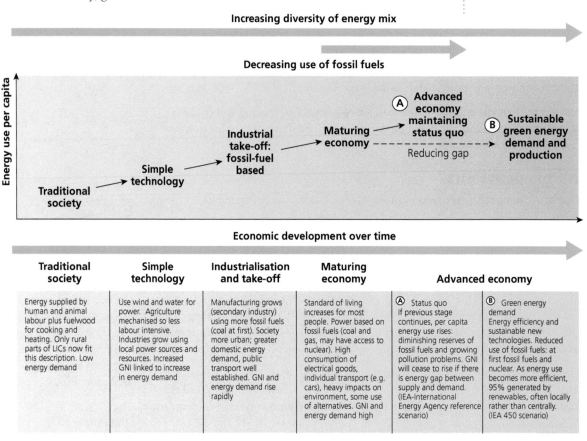

Figure 22 An energy transition model

Sweden's economy is very energy intensive due to its broad manufacturing base and high living standards. Sweden has one of the lowest carbon economies of all OECD countries: 92 per cent of its electricity is generated from renewables, predominantly hydropower and biomass. Greater efficiency has reduced the demand per capita and lowered costs (in 2000 usage of electricity was 15.7 mWh per capita, falling to 13.5 mWh per capita by 2014). Sweden aims to achieve a fossil-fuel-independent fleet by 2030. However, there are concerns about the disposal of nuclear waste and the need to adopt clean energy technologies for its industrial sector.

India's energy consumption grew by over 300 per cent between 1973 and 2005. The contribution of fuelwood/biomass to its energy supply has declined and India has embarked on a programme of developing renewables, especially solar, wind and hydropower, as well as a nuclear programme. It lacks fossil fuel resources, except for coal (p. 30).

Botswana's energy consumption is growing from a low base. Traditionally fuelwood was the main source of energy. Due to its semi-arid climate, low wind speeds (high pressure dominates) and land-locked location, the greatest potential is from solar power in the form of renewable microgeneration projects that can power rural communities. Technology transfer and aid programmes are important in supporting these projects. With reserves of 200 billion tonnes, coal is another important resource.

Economic and political factors affecting world energy prices and mix

The current global energy mix exhibits an over-dependency on fossil fuels. Global markets for oil, natural gas and renewable energy are complex and constantly changing.

Oil prices are particularly volatile. They quadrupled during the oil crisis of 1973–74 following the Arab-Israeli war, when OPEC cut oil supplies dramatically. Another oil crisis occurred in 1979, when a Shia Muslim theocracy overthrew the Shah of Iran. This was followed by the Iran-Iraq war and the two gulf wars in Kuwait (1990–1) and Iraq (2003–10). The effect of these major crises was a sudden increase in oil price and, in the case of the Iranian Revolution, the start of a rise in non-OPEC production and a decline in OPEC production. Instability in the Middle East and the increasing production of oil from shale, particularly in the USA, continues to affect oil prices.

China is the world's leading consumer of fossil fuels, although its economy and energy mix are changing from heavy industry fuelled by coal to a service-based economy with a more varied energy mix (p. 45). High levels of air pollution in cities have been the catalyst for China's ambition for a cleaner energy mix. China now accounts for 60 per cent of global solar cell manufacturing capacity.

Japan's Fukushima nuclear plant meltdown, caused by an earthquake-generated tsunami in 2011, prompted many countries to re-evaluate their energy policies. Fukushima had a particularly drastic effect on policy in Germany, where there are plans to close all the country's nuclear reactors by 2022. France remains pronuclear, but plans to reduce its nuclear share in order to diversify the energy portfolio.

> **Exam tip**
>
> Consider the different specialised concepts you could apply to an exam question on factors affecting world energy prices and energy mix. Where could you integrate the specialised concepts of causality, globalisation, mitigation, resilience, risk and sustainability into your discussion?

The need for sustainable solutions

Policies for demand reduction and increased energy efficiency

HICs can reduce their demand for energy through the installation of smart meters, which show people how much energy they are using, encouraging them to change their usage patterns. Walking or cycling instead of using a car is a direct reduction. Providing high-quality, affordable public transport to cut car usage (e.g. bus rapid transport systems (BRT), such as that used in Bogota, Colombia), congestion charging (e.g. London), and procurement of low-carbon vehicles, such as hydrogen-powered buses or electric cars (e.g. Reykjavik) also achieve reduction. Modern telecommunications (e.g. teleworking, video conferencing and Skype) can make travelling to attend meetings redundant. For LICs, reducing energy demand is more problematic because most people in LICs use less energy anyway.

> HICs stands for high-income countries. The World Bank classifies countries into four categories: low-income countries (LICs); lower middle-income countries (LMICs); upper middle-income countries (UMICs); and HICs.

It is estimated that carbon emissions could be stabilised just through greater efficiency. Research in China concluded that, from Shanghai southwards, light-coloured roofs reduce the need for air conditioning by reflecting more sunlight, lowering annual energy usage and costs as well as annual emissions of carbon dioxide, nitrous oxide and sulphur dioxide. As China takes steps to improve its air quality, there will be more benefits from cooler, lighter surfaces because more sunlight will be striking buildings and lighter surfaces reflect more sunlight, reducing the need for air conditioning.

Singapore aims to have 80 per cent of its buildings achieving a 'Green Mark' standard by 2030. In a housing development in the Punggol area of the city the buildings face towards the wind and away from the sun, natural ventilation is favoured over air-conditioning, rooftops collect rainwater and protect against the sun, and plants insulate against the heat.

In the EU, new cars in 2040 are likely to be around 70 per cent more efficient than in 2000. Efficiency measures in transport represent over a third of urban mitigation potential in the period to 2050.

Clean technologies for fossil fuels and transport technologies

Radical technologies offer a solution to reducing carbon emissions whilst maintaining a fossil fuel-based economy. As coal is an abundant and cheap energy source, one alternative is to capture the CO_2 released by burning it (or oil and gas) and storing or burying it, for example underground (Figure 23). However, carbon capture is expensive because of the complex technology involved, and the success of trapping carbon underground indefinitely is not yet proven.

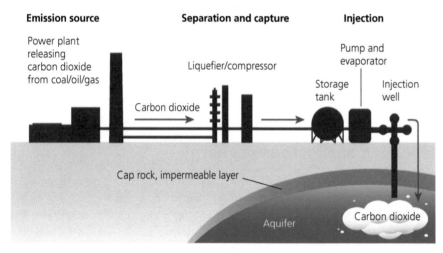

Figure 23 How carbon capture works

Carbon sequestration is the natural capture and storage of CO_2 from the atmosphere by physical or biological processes such as photosynthesis. Gasification involves removing CO_2 from natural gas, either at the point of production or at energy facilities from which gas is distributed.

Hydrogen fuel cells are a promising technology for use as a power source for electric vehicles. Hydrogen does not occur naturally as a gas, because is always combined with other elements. Hydrogen is high in energy, and burning hydrogen produces almost no pollution. A fuel cell combines hydrogen and oxygen to produce electricity, heat, and water. Fuel cells are often compared to batteries. Both convert the energy produced by a chemical reaction into usable electric power.

Knowledge check 33

Define the term 'carbon neutral'.

Sustainability of alternative energy sources

At present, most energy is carbon based. Suitable alternatives are essential, with increased use of renewables offering a major pathway to a more sustainable energy future. It is argued that one of the key problems with wind, solar and tidal energy — the unpredictability of supply — could be overcome by 'storing' it in porous rocks as compressed air. When energy is required, the 'stored' air can be allowed to escape to drive turbines that regenerate electricity.

Wind

Wind energy is directly related to solar activity, which causes differences in temperature and therefore atmospheric pressure, generating winds. Wind energy has great potential where winds are strong and particularly in winter, when the demand for energy is at its highest. However, as energy can only be generated when the wind blows, this form of alternative energy is location specific (higher altitudes, exposed locations) and is often heavily contested in the scenic areas such as mid-Wales, where many wind farms are located. In 2017 44 per cent of Denmark's electricity was supplied by wind (including offshore wind farms).

Hydropower

Hydropower is one of the most mature of the alternative sources of energy. Although it generates a cheap form of electricity compared with other renewables, the initial cost of infrastructure can be high (p. 69), and construction often destroys local river ecosystems (e.g. the Yangtze river dolphin may now be extinct) and communities. The methane released from the resultant decaying vegetation is a powerful greenhouse gas. Venezuela relies heavily on hydroelectric power from the Guri Dam, built in the 1970s/80s. However, hydropower met less than 25 per cent of Venezuela's energy demands in 2014 due to drought conditions associated with El Niño.

Solar

After hydropower and wind, solar is the third most important renewable energy source in terms of globally installed capacity. It is pollution free, efficient, requires little maintenance and in theory provides a limitless supply of energy. Only recently has photovoltaic technology become cheaper compared with fossil fuels, and solar energy varies spatially and temporally (insolation is limited at high latitudes and in mid-latitudes in winter). Moreover, many suitable sites such as the Sahara Desert are great distances from major centres of population, and require the development of expensive transmission lines. Singapore produces 15 per cent of its energy from solar energy, with the potential to increase this to 30 per cent.

Geothermal

Geothermal energy provides a constant supply of energy, both in terms of hot water for space heating and electricity generation from steam. It is extremely competitive in suitable areas of tectonic activity such as Iceland and New Zealand, but the range of locations where geothermal energy can be exploited cost effectively is limited, given the current technology. Reykjavik, Iceland meets virtually all its heating needs with geothermal energy, with 99.9 per cent of the city's heating provided by geothermal district heating.

Exam tip

When discussing the sustainability of energy sources, ensure that you define the term 'sustainability', to provide you with a benchmark with which to assess how sustainable alternative energy sources are.

Knowledge check 34

Which two renewable sources of energy together generate 100 per cent of Iceland's electricity?

Biomass

Organic material of biological origin is easy to source and can reduce dependence on fossil fuels, but can lead to rainforest destruction and compromise food production. Biofuels such as maize and sugar cane may require irrigation, and the production, harvesting and transport of biofuels create a carbon footprint. Vaxjo, Sweden uses forestry waste transformed to biomass energy to supply 40 per cent of its electricity and 80 per cent of heating. Curitiba, Brazil uses biodiesel-only buses.

Algae can grow in salt water and on land unsuitable for crops, so a successful algae-based biofuel could provide the world with more energy without posing a challenge to global fresh water and food supplies.

Wave, tidal, ocean

Much of the energy potential of oceans is yet to be developed. The proposed Swansea Bay tidal lagoon has the potential to supply energy equivalent to 90 per cent of Swansea Bay's annual domestic electricity use (to over 155,000 homes). However, at a cost of £85 million, there are problems with funding this form of energy, and there is concern over its environmental effects.

Nuclear

Nuclear is a major supplier worldwide but generates debate. It produces long-lasting, dangerous waste, it could fall into the hands of terrorists, and it can be used to produce nuclear weapons. However, there have been few leaks, and containment is improving. Otherwise, nuclear produces clean and plentiful energy.

Knowledge check 35

Identify countries at different levels of development that have, or are developing, nuclear energy programmes.

Summary

- Energy resources are classified as either non-renewable or renewable, and are unevenly distributed.
- Alternatives to conventional fossil fuel sources include shale gas, tar sands, deepwater oil, renewables and nuclear energy.
- Physical factors determine the supply of energy. Geological factors influence the distribution of fossil fuels and geothermal energy. Solar and wind energy are influenced by climate, hydropower by climate and relief, and wave/tidal power and biofuels by favourable locational conditions.
- Demand for energy is driven by a combination of population growth, economic development, increased living standards and technological changes.
- The imbalance between the supply and demand for oil is managed through transfers, storage and pricing. Players responsible for managing oil and gas resources include MNCs, OPEC and national governments.
- There are environmental, political, technological and economic problems associated with fossil fuels and other forms of energy.
- The energy mix varies according to levels of development, and can be examined at the local, national and global scales.
- Sustainable solutions for meeting energy demand include reduction policies, improvements in energy efficiency, technologies that reduce carbon emissions and alternative energy sources.

■ Weather and climate

Global controls on climate

Structure of the atmosphere

The atmosphere can be divided into four main layers:

■ Troposphere — most weather processes take place in this lowest layer, which contains most of the atmosphere's mass, water vapour and dust. It varies between 8 km and 12 km in height. The lower layers of the troposphere are largely heated from below, with temperatures dropping by 6.5°C approximately every 1000 m as the atmosphere thins. The tropopause separates the troposphere from the stratosphere.

■ Stratosphere — this lacks dust and water vapour and is largely stable. An increase in temperature with height results from the absorption of solar radiation, particularly ultraviolet wavelengths, by ozone molecules, which form a layer 20–30 km above the Earth.

■ Mesosphere — here temperatures again fall with height because of the decreasing density of the atmosphere, and therefore its inability to absorb energy.

■ Thermosphere — this lies at a height of above 90 km. Temperature again increases with altitude, due to absorption of radiation at ultraviolet wavelengths.

The atmospheric heat budget (Figure 36 on page 59 of *WJEC/Eduqas A-level Geography Student Guide 4: Water and carbon cycles*) is the balance between incoming solar radiation (insolation) and outgoing radiation from the planet. The sun is the source of energy that drives the atmospheric engine.

The amount of solar energy received varies with latitude (Figure 24). The tropics have an energy surplus because they gain more from insolation than is lost by radiation. The higher temperate and polar latitudes have an energy deficiency, losing more by radiation than is gained by insolation. This imbalance in energy distribution sets up a transfer of heat energy via winds and ocean currents from the tropics to higher latitudes.

> **Knowledge check 36**
>
> What is the effect of an increase in greenhouse gases on the atmospheric heat budget?

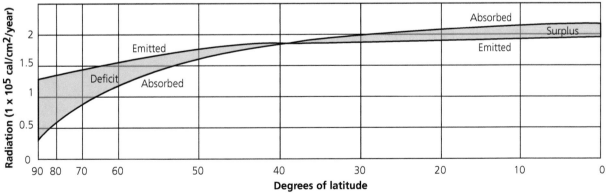

Figure 24 Heat budget changes with latitude

Processes of global atmospheric circulation

This global transfer of energy is the basis of global atmospheric circulations, which give rise to the low- and high-pressure belts and the planetary wind systems associated with the Earth's three major convection cells: the Hadley, Ferrel and polar cells. These make up the tricellular model that controls atmospheric movements and the redistribution of heat energy.

As more has been learned about global circulation, the basic tricellular model has been modified (Figure 25). The troposphere thins towards the poles and the tropopause is broken into three distinct zones. At these points, the powerful eastward-moving, high-altitude winds called jet streams occur. In the Hadley cell, for example, the subtropical jet stream contributes to movement of air along an east–west axis (Walker cell).

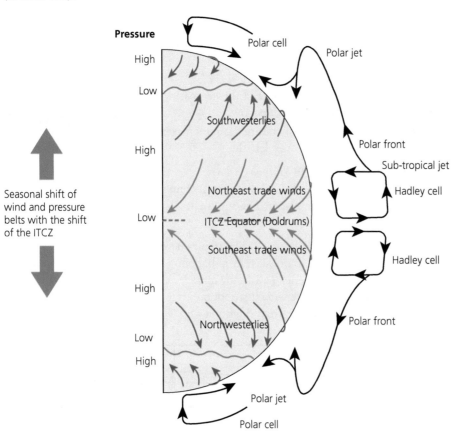

Figure 25 Pressure belts and associated wind systems

High- and low-pressure belts, and the impact of oceanic circulation, continentality and altitude on climate

The pattern of the planetary surface winds is shown in Figure 25. Winds blow from high pressure to low pressure. However, as a result of the Coriolis force caused by the Earth's rotation, winds are deflected to the east in the northern hemisphere and to the west in the southern hemisphere.

In addition to differences in solar energy received between low and high latitudes, oceanic circulation, continentality and altitude also exert an influence on climate.

Ocean currents

Ocean currents are set in motion by the wind blowing across the surface of the oceans. The pattern of world ocean currents is the result of the tilt of the Earth, its rotation and the distribution of the land masses.

Oceanic circulation follows a largely circular route (gyre) in each ocean basin (the exception being the circular polar current that flows around Antarctica). Ocean currents move clockwise in the northern hemisphere and anticlockwise in the southern hemisphere.

One example of a warm current is the North Atlantic Drift or Gulf Stream, which flows off the west coast of the UK. One example of a cold current is the Labrador Current, which flows off the east coast of Canada and the USA in the north Atlantic Ocean.

Continentality

A major influence on regional climate is distance from the sea, known as continentality. The land surfaces of continents are solid and respond readily to solar radiation (as they have a lower **specific heat capacity** than the sea), heating up quickly during the summer and cooling down rapidly in winter, resulting in a greater annual temperature range. Oceans heat up more gradually during the summer but retain this heat during the winter, leading to a smaller annual temperature range. This means that places far inland have a greater annual temperature range than those on the coast.

Altitude

Temperatures decrease with altitude. Globally the reduction of temperature with altitude within the troposphere, known as the environmental lapse rate (ELR), occurs at a rate of 6.5°C per 1000 m. If dry air is forced to rise over mountains such as the Rockies, the consequent expansion in volume causes a temperature fall of 10°C per 1000 m (the dry adiabatic lapse rate (DALR)). If the air becomes saturated and condensation occurs there is a release of latent heat, which reduces the rate of cooling to 5°C per 1000 m (the saturated adiabatic lapse rate (SALR)).

Exam tip

Use diagrams to help you explain difficult meteorological concepts more easily.

Knowledge check 37

Which currents are associated with positive latitudinal temperature anomalies?

Knowledge check 38

Why are some cold currents associated with coastal deserts?

Specific heat capacity is the heat required to raise the temperature of the unit mass of a given substance by a given amount (usually one degree).

Knowledge check 39

Which of the following locations, Vancouver (on the Pacific coast) or Winnipeg (in the Canadian interior), both found at 49°N, will experience the greatest annual temperature range?

Content Guidance

World's major climate types

The world's major climate types are shown in Figure 26.

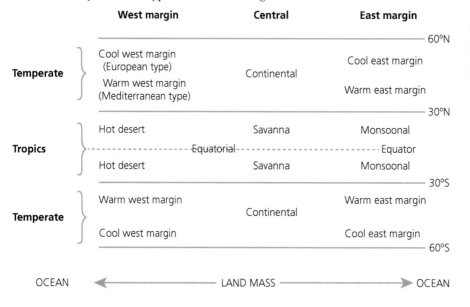

Figure 26 The distribution of the main climate types for the tropical and temperate latitudinal belts

The main influences on the climates of tropical regions are:

- the overhead or near-overhead position of the sun giving high insolation throughout the year
- the position and seasonal movement of the **ITCZ**, together with the wind systems of the tropical pressure belts
- the path of the upper jet streams affecting the paths of low-pressure systems

Additional influences include oceanic circulation, continentality and altitude.

One tropical climate type is the savanna climate type, located 5–20° latitude either side of the Equatorial belt. The savanna climate type is characterised by high temperatures of 35–25°C all year, because insolation is high. It is also distinguished by having a hot, wet season and a marginally cooler, dry season. Humidity is highest in the wet season, but evaporation rates remain high during the dry season. Rainfall occurrence is linked with the movement of the ITCZ towards the tropic in association with the apparent movement of the sun's position overhead. During the hot season, as this occurs, low pressure prevails with moist in-blowing winds and rising air currents leading to convection rainfall. Rainfall amounts are most reliable towards the Equatorial latitudes, where they average 800 mm a year, but become less reliable towards the hot desert margins, where they average 300–400 mm annually.

The cooler, dry season in the savanna belt occurs at the time when high pressure and dry, out-blowing winds prevail; this is when the overhead sun and the ITCZ move

Exam tip

The specification requires you to make detailed reference to the monsoon climate and the UK's climate (west margin European maritime type), so knowledge and understanding of these two will provide the basis for answering any questions on major climate types.

Exam tip

If one of your option themes is *Development in an African context*, knowledge of the savanna climate type will assist with an understanding of physical factors influencing the development of desertification.

ITCZ (intertropical convergence zone) is a zone of low atmospheric pressure and ascending air formed where the trade winds converge and convection results from thermal heating. The ITCZ migrates northwards and southwards of the Equator with the seasons. It is sometimes referred to as the ITD (the intertropical discontinuity).

away to extend beyond the equator towards the other tropic. The persistence of high pressure over the outer margins of the savanna may prevent the ITCZ and in-blowing moist winds from extending into these areas, leading to drought conditions.

> **Exam tip**
>
> Learn climatic data for your chosen climate type to exemplify its characteristic precipitation amount and distribution, and variations in temperature at a diurnal and seasonal level. Use these figures in your answers to make them more rigorous and detailed, and to strengthen your AO1 marks.

For temperate regions, see 'Climate and weather of the UK' (p. 81).

Seasonal variations in the position of the ITCZ, heat equator and wind and pressure belts

The combined effects of the Earth's axial tilt and its orbit around the sun results in the seasons. The sun's overhead position shifts during the year from the equator (where it is overhead at the Equinox on 21 March*) to the Tropic of Cancer (where it is overhead on 21 June) and returns southward to the equator (where it is overhead at the Equinox on 21 September) and to the Tropic of Capricorn (where it is overhead on 21 December).

*The exact dates of each equinox and solstice can vary from the 20th to the 23rd of the month.

The relative movement of the sun's overhead position shifts the point of maximum insolation seasonally (the **heat equator**). This influences the latitudinal movement of the pressure belts and wind systems, notably the ITCZ. The ITCZ is the point at which the two Hadley cells meet and where the trade winds converge, and moist tropical air rises to give rain. The movement of the ITCZ is key to the seasonal incidence of rainfall in the tropics. Over the oceans the ITCZ does not shift beyond 5° either side of the equator, but over land the latitudinal seasonal shift is greater to the north of the equator than to the south.

Monsoon climate

Monsoon climates occur mainly on the eastern side of continental landmasses in the tropics, extending across approximately 5–20° of latitude.

The monsoon climate is marked by a distinct hot, wet season and a cooler dry season, determined by the annual movement of the ITCZ between the tropics, and the consequent movement of pressure belts and seasonal reversal of winds. The monsoon climate regime is most clearly seen in the Indian subcontinent, because of the size of the landmass and its relief, but exists in other regions north and south of the equator on the eastern edge of continents, for example in east Africa.

The wet monsoon season (Figure 27) occurs with the movement of the ITCZ into the region. This brings an area of low pressure and draws in hot, moist winds from the ocean. Rainfall is increased by orographic uplift, where these moist winds are drawn over uplands, for example the Western Ghats in India. Temperatures are high,

> **Knowledge check 40**
>
> What is convection rainfall? Name two other types of rainfall (you need to recognise synoptic links with the water cycle).

The **heat equator** (or thermal equator) moves north and south seasonally with the changing position of the overhead Sun. It is the continuous area on the globe that has the highest surface temperatures.

averaging 30°C. Humidity is also very high, with average rainfall around 2000 mm, decreasing with distance inland. Cyclones and hurricanes are frequent towards the end of the rainy season.

Knowledge check 41

Explain why the seasonal shift of the ITCZ is greater over land than over the oceans and greater to the north of the equator than to the south.

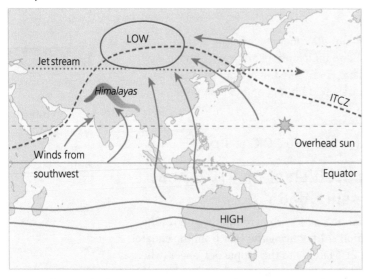

Figure 27 The wet monsoon season (June to October)

The cooler dry season (Figure 28) coincides with the extension of continental high pressure as the ITCZ moves back towards the equator and across into the tropics beyond. With high pressure dominating, there is air subsidence and the out-blowing winds are dry. Temperatures remain relatively high, at 25°C in lowland areas, and evaporation rates are also high. The weather is much more severe in mountain areas.

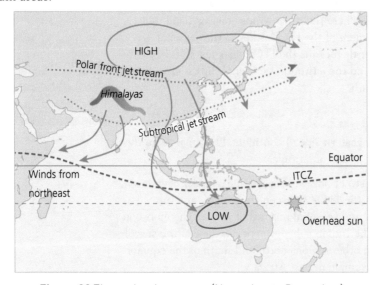

Figure 28 The cooler dry season (November to December)

Climate and weather of the UK

Characteristics of the UK's climate

The UK has a cool temperate western margin climate (maritime).

The UK's climate is characterised by relatively mild temperatures (average seasonal range 5–20°C), along with high humidity and precipitation (averaging 600 mm) throughout the year. However, precipitation totals are significantly higher over upland areas in the face of prevailing moist, westerly winds coming off the ocean, for example in the Cambrian Mountains of Wales. Conversely, precipitation totals are low in rain-shadow areas, such as lowland East Anglia.

The temperatures and precipitation figures are mainly influenced by the mid-latitude position, the low-pressure belt and the mild westerly prevailing winds. The latter are warmed by warm currents, such as the Gulf Stream, on the west margins of landmasses.

Air masses and their influence

The weather is strongly influenced by the position of the **polar front** (Figure 25, p. 76), the associated jet stream, and the passage of westerly-moving depressions along the front, with intervening spells of anticyclonic conditions. These are linked to the position and extent of the main air masses influencing the continental west margins in mid latitudes: the polar continental, polar maritime, Arctic maritime, tropical maritime and tropical continental air masses. The interaction between these air masses, together with the associated upper jet stream and Rossby waves, influence the occurrence and development of depressions along the polar front.

Persistence of one of the continental air masses across these western margins can bring long spells of dry summer weather, while in winter 'anticyclonic gloom' conditions may occur. In contrast, the passage over the area of a deep, fast-moving depression can bring storm conditions with gale-force winds and heavy rainfall.

The **polar front** is the boundary between the polar cell and the Ferrel cell, at around the 60° latitude in each hemisphere.

The jet stream

A belt of upper-air, westerly winds occurs vertically above the polar front, caused by marked temperature contrasts between tropical and polar air. Temperature contrasts at the surface produce marked pressure gradients in the upper air, which in turn lead to extremely strong winds. These upper-air winds follow meandering paths called Rossby waves. The amplitude and sinuosity (pattern) of these Rossby waves varies seasonally, with four to six waves in summer and usually three in winter, in a continuous belt around the globe in mid-latitudes. Within the Rossby waves are jet streams, narrow bands of extremely fast-moving air, which decelerate as they turn away from the poles and accelerate as they move poleward.

In the mid latitudes of the northern hemisphere, Rossby waves are associated with the formation of depressions and anticyclones. The jet stream travels west to east within the Rossby waves, but in a wave-like movement, so when it flows towards the equator it slows down and piles up, causing convergence. Convergence in the upper air causes a downflow to the ground, creating high-pressure systems at ground level. As air travels poleward, it speeds up and diverges ahead of the next trough. Divergence in the upper air causes low-pressure systems at ground level. At times the waves are few and shallow, giving a high **zonal index** and a succession of low-pressure systems. At other times the flow becomes more pronounced, giving a low zonal index and causing the formation of blocking, high-pressure systems.

Extreme weather events

Recent and cyclic climate change

El Niño and La Niña are terms for cyclical climatic events originating in the tropical Pacific, which are thought to be becoming more frequent and intense because of global warming.

These episodes alternate in an irregular cycle called the ENSO (El Niño southern oscillation). 'Southern oscillation' refers to atmospheric pressure changes between the east and west tropical Pacific that reflect the close interaction between the atmosphere and ocean. They are associated with widespread changes in the climate system, such as heavy rainfall, flooding and extreme drought, which last several months, and can lead to significant socio-economic impacts affecting infrastructure, agriculture, health and energy sectors.

Changing vulnerability of populations to weather and climatic hazards

The vulnerability of populations to climatic variability and to weather and climatic hazards varies both spatially and sectorally. The most vulnerable regions are likely to be:

- the Arctic, due to the impacts of high rates of projected warming (p. 88) on ecosystems
- sub-Saharan African countries, such as Somalia, due to their low **adaptive capacity**
- the deltas of many Asian rivers, such as the Ganges-Brahmaputra-Meghna Delta (Asia's largest, and the world's most populated delta), because of their large populations and exposure to river flooding, sea-level rise and storm surges, as well as being subject to stresses imposed by human modification of catchment areas and delta plain land use
- small, low-lying islands, such as Kiribati and Tuvalu in the Pacific, because of their exposure to sea-level rise and storm surges

The most vulnerable sectors are likely to be:

- health, in areas of low adaptive capacity
- agriculture, because of reduced water availability
- water resources in mid-latitudes and the dry tropics, due to decreases in rainfall and higher evapotranspiration rates
- ecosystems, particularly mangrove forests, salt marshes, coral reefs and tropical rainforests
- low-lying coastal regions, because of their exposure to sea-level rise and extreme weather events

The **zonal index** is a measure of strength of the mid-latitude westerlies, usually expressed as the horizontal pressure difference between 35°N and 55°N.

Knowledge check 42

Give one reason why projected temperature increases for the Arctic are higher than the global average.

Adaptive capacity is the ability of governments, groups of people and individuals to adjust to changes and to manage the consequences of change.

Impacts and management of climatic hazards

Hazards associated with low-pressure systems

In the tropics, low-pressure hazards are tropical storms and cyclones, with torrential rain and high winds. These **climatic hazard** conditions, usually created towards the end of the hot season (August–November in the northern hemisphere), are generated in exceptionally deep, fast-moving depressions over oceans off the east margins of continents, in the tropics and subtropics. These depressions travel westwards due to the Coriolis force and grow in size, fed by the ocean's heat and water evaporating from the ocean surface. They trigger the secondary hazards of flooding, storm surges and sea incursions, landslides, mudflows and windborne debris.

In the temperate region, low-pressure hazards include severe storms, heavy rainfall or snowfall and gale-force winds. These conditions are generated in exceptionally deep and fast-moving depressions, which are most likely to occur in autumn and spring along the polar front. They trigger the secondary hazards of flooding, sea incursions (especially where the deep depression coincides with very high tides), landslides and windborne debris.

During the hurricane season of 2017, a succession of hurricanes — Harvey (category 4 on the **Saffir-Simpson scale**, with 120 mph winds affecting Texas and Louisiana), Irma (category 4, with 130 mph winds affecting Florida, Georgia and South Carolina and damaging 95 per cent of the buildings in Barbuda), Maria (category 4, with 155 mph winds) and Nate — impacted the Caribbean and the Gulf Coast of the USA. Table 8 summarises the impacts of hurricane Maria on Puerto Rico.

> A **climatic hazard** is an extreme climatic/weather event causing harm and damage to people, property, infrastructure and land uses.

> The **Saffir-Simpson scale** is a 1–5 hurricane wind scale, which estimates potential damage to property and loss of life based on a hurricane's sustained wind speed. Hurricanes reaching category 3 and higher are considered major hurricanes.

Table 8 The impacts of hurricane Maria, September 2017

Hurricane Maria	Economic impacts	Social impacts	Environmental impacts
Maria hit the US territory of Puerto Rico in the middle of September. It was the first category 4 hurricane to directly impact the island in 85 years	In the immediate aftermath, nearly all of Puerto Rico's 3.4 million US citizens were without power The main elements of Puerto Rico's economy — agriculture, fisheries and tourism — were devastated by the storm Puerto Rican authorities have requested $94 billion to cover damages	The official death toll is 55, but almost 500 deaths were recorded by funeral homes that could have been caused by the storm 13 people are unaccounted for Hospitals and other public buildings remain badly damaged Puerto Ricans lack sustainable sources of food, power, and income People are migrating to settle in mainland US states	Massive mudslides Suspended matter floating in the water limits the amount of sunlight reaching marine habitats, reducing growth and recovery Destruction of extensive seagrass meadows that support lobster, shrimp, conch and finfish fisheries, stabilise sediments and protect the white sand beaches that attract tourists

Hazards associated with high-pressure systems

In tropical climates, high-pressure hazards are low rainfall, high evaporation rates and drought (Table 9). These trigger the secondary hazards of falling water tables, loss of vegetation, wildfires, soil erosion and associated desertification. These hazards are associated with anticyclonic conditions, which are due to the continued persistence of subtropical high pressure over continental areas. This limits the ITCZ zone to lower latitudes (nearer the equator) than is normal for the time of the year.

In temperate climates, high-pressure hazards include drought in summer and frost and fog in winter. They may trigger secondary hazards in summer (falling water tables and loss of vegetation) and in winter (temperature inversion, with air pollution intensifying the fog conditions). These conditions are associated with persistent stationary anticyclones, which in summer are usually associated with the extension into higher latitudes of subtropical high pressure. In winter the conditions are usually associated with the extension of the continental high pressure towards the coastal margin of the landmasses.

Exam tip

Try to provide examples of the impacts of hazards associated with recent climatic events to make your answers as topical as possible.

Table 9 The impacts of the Kenyan drought in 2014–18

Causes	The persistence of high pressure over the outer margins of the savanna prevented the ITCZ and in-blowing moist winds from extending into these areas, leading to drought conditions Global warming was a further contributory factor, exacerbated by people's misuse of their environment
Economic impacts	Maize production in the coastal areas decreased by 99 per cent compared with the long-term average Pastoralist communities in the arid and semi-arid lands reported large numbers of animal deaths
Social impacts	More than 3.4 million Kenyans were severely food insecure as of May 2017 An estimated 3 million people lacked access to clean water Conflicts over scare resources reduced the ability of communities to cope Migration, which posed risks for women and children Multiple disease outbreaks, including cholera/diarrhoea and measles People travelling further to access water; for example in Baringo, the household walk to access water was three times longer than normal A decline in school attendance and school participation, and rising dropout rates
Environmental impacts	Severe drought dried up water resources in half of Kenya's 47 counties Soil erosion Surface soil became wind-blown and covered emerging vegetation, affecting grazing Surface wash (when rains did come) produced deep gulleys Soil filled these gulleys, which then flooded during occasional downpours

Strategies to manage climatic hazards

Strategies to reduce the impact of hazards associated with low-pressure and high-pressure systems include monitoring, prediction and warning of future hazards, immediate response to lessen the impact once the hazard has occurred, and long-term planning.

Increasingly, strategies involve the use of technology:

- Modifying human vulnerability to hazard risk through prediction and warning, community preparedness and land use planning. Technological advances include: weather satellites, ocean buoys, radar and computer modelling to forecast, track and predict climatic hazards associated with low-pressure storms or high-pressure droughts; the use of GIS to develop risk mapping; the use of GM technology to develop drought-resistant or salt-tolerant crops to ensure improved food supplies during a period of drought.
- Modifying the hazard event through environmental control and hazard-resistant design. Technological advances include: cloud seeding, building design, hazard proofing and the development of various defence systems, such as flood walls and control dams.
- Modifying the loss through aid and insurance.

In the case of hurricane Katrina, which struck the Gulf coast of the USA in 2005, hurricane warnings were given, and emergency services were in place, but the strategies were not successful due to the failure of the levées to protect the city of New Orleans and due to the slow response of the US Federal Emergency Management Agency (FEMA).

LICs are usually less prepared and often depend on aid from HICs. Other strategies used in poorer countries include building hurricane shelters, strengthening and raising embankments, planting mangrove trees to absorb storm surges and educating people about the risk.

In the case of drought in Kenya, sustainable solutions include water saving, water reuse and water treatment. Small-scale adaptive strategies include stone stripes, rainwater harvesting, more efficient irrigation and sand dams (p. 58). Regional- and national-scale strategies include replanting schemes, diversification (e.g. into beekeeping) and developing opportunities for small business enterprises, particularly for women.

Impacts of human activities on the atmosphere at local and regional scales

Impacts of urban areas on temperature, wind, precipitation and humidity

Urban areas replace existing microclimates with new ones. Table 10 summarises the causes and characteristics of changes to microclimate variables caused by urbanisation.

Exam tip

When evaluating the impacts of climatic hazards, think about whether the short-term impacts are greater than the long-term impacts, or whether the economic, social or environmental effects are the most important. Note that many of the impacts are interrelated.

Exam tip

You should show knowledge and understanding of the climatic causes of, and the weather associated with, low- and high-pressure systems, together with the human circumstances that constitute the hazard.

Table 10 Average changes to microclimates caused by urbanisation

Microclimate variable	Characteristics	Causes
Temperature	In most mid-latitude European and North American cities, average minimum winter temperatures are 1–2°C higher than in rural environments	Urban environments are made up of heat-retaining materials with better radiation-absorbing properties and lower albedo rates (p. 56) Heat is released from buildings, vehicles and industry Due to the absence of vegetation, less solar energy is used up in evapotranspiration, so more is available to heat the surface
Wind speed	Annual mean wind speeds are around 20 per cent lower and the frequency of extreme gusts around 15 per cent lower than in rural environments The incidence of calms is around 20 per cent higher	The frictional drag of the urban landscape reduces wind speeds Calm conditions contribute to a higher incidence of fogs In some cities, tall buildings funnel the wind, increasing wind speeds
Precipitation	Total precipitation is 5–30 per cent higher The number of rain days is 10 per cent more and snow days are 14 per cent less than in rural environments	The greater amount of dust in urban areas increases the concentration of hygroscopic particles that provide condensation nuclei Higher temperatures result in increased convectional rainfall but less snow
Relative humidity	Relative humidity is 20 per cent less in winter and 8–10 per cent less in summer than in rural environments	A lack of water bodies, less vegetation and higher temperatures (at higher temperatures, the atmosphere has a greater capacity to hold water in its vaporous state than at lower temperatures) result in lower relative humidity

Relative humidity is the amount of water vapour actually present in the air compared with the greatest amount it could hold if it was fully saturated.

The population of Barrow in Alaska grew from 300 in 1990 to 4600 in 2017. Research carried out during the winter of 2001/2002 reported that, on average, Barrow was 2.2°C warmer than the surrounding rural area, creating an **urban heat island**. A maximum difference of 6°C recorded in January and February reflected higher energy usage for the heating of residential and commercial buildings.

Urban heat island refers to the distinctive climate associated with an urban area, particularly in terms of temperature. Human activities result in a warming of the urban area compared with the surrounding rural area.

Exam tip

Consider factors that influence the intensity and shape of the urban heat island. Which factors other than population size, population density, urban sprawl and nature of economic activity will influence it?

Impacts of urban areas on air quality

Urban areas can create particulate pollution, photochemical smog and acid rain. Table 11 explains the atmospheric composition of average urban areas.

Table 11 Atmospheric composition of urban areas — causes and impacts

Air quality	Causes	Impacts
Particulate pollution	Road traffic emissions, particularly from diesel vehicles Particulates are also emitted from power generation, commercial and residential combustion and some non-combustion processes (e.g. quarrying)	Particulates reduce the amount of incoming solar radiation (by 15–20 per cent) and increase the number of condensation nuclei, leading to fewer hours of sunshine and more rain, respectively
Acid rain	Rain contaminated by chemicals, notably sulphur dioxide through fossil fuel combustion (principally power stations), conversion of wood pulp to paper, manufacture of sulphuric acid, smelting and the incineration of refuse	Destruction of buildings as a result of carbonation, especially a problem on ornate, historic buildings made from limestone (e.g. Parthenon, Taj Mahal) Acid rain also affects human health (airways and lung function)
Photochemical smog	Formed where intense sunlight, high concentrations of volatile organic compounds and nitrogen oxides from car traffic (largely diesel exhausts) combine to form a poisonous smog containing ozone, liquid particulates and other pollutants, leading to a brown haze of nitrogen dioxide	Irreversible damage on the lungs and heart Even short-term exposure to photochemical smog can have ill effects on the young and the elderly High levels of smog trigger asthma attacks because the smog causes increased sensitivity to allergens, which are triggers for asthma

Strategies to reduce the impact of human activity on urban climates and air quality

Mitigation can be based on technological solutions, political solutions or attitudinal fixes. Evidence suggests that this can be successful, with developed countries leading the way.

- Technological solutions include:
  - controlling particulates from cars (catalytic converters)
  - encouraging a shift from diesel- and petrol-driven cars to electric cars
  - decreasing harmful emissions from power stations and industrial plants (scrubbers decrease harmful emissions)
  - developing alternative energy sources
- Political solutions — for example, smog prevention can occur through legislation (Clean Air Acts). Cities increasingly use systems to prevent vehicles entering (congestion charging in London).
- Attitudinal fixes include educating people to purchase electric/hybrid fuel cars, supported by taxation perks, or to change their energy use.

Knowledge check 43

Why is acid rain often described as 'someone else's problem'?

Knowledge check 44

Why is urban air quality poorer during early morning rush hours in winter?

People, climate and the future

Global impact of anthropogenic climate change

Although predictions are that anthropogenic climate change will affect all climate belts, IPCC assessments report evidence of pronounced changes in Arctic regions north of 65°N. The global mean temperature rise measured at the surface between 1900 and 2005 has increased by 0.74°C, but polar warming has been over twice that rate, due to positive **feedback**. The expected rise in temperature will have further impacts on other climatic elements, such as pressure and winds.

Effects on the Arctic region associated with climate change include the following:
- The boundary between the taiga (boreal) forest and the tundra has advanced northwards in response to higher temperatures.
- The length of the growing season has increased (3 days per decade in Alaska).
- Research has provided evidence for the destabilisation of the western Arctic high-pressure zone (the 'Beaufort high'). Warmer temperatures have resulted in thinner and less extensive sea ice, allowing for more oceanic heat to be transferred to the atmosphere, providing an additional energy source for storms. During the winter of 2017, low-pressure systems moved into the western Arctic, causing the collapse of the Beaufort high, which had never happened before.

The atmospheric tipping point

A 2018 IPCC report concluded that the world faces environmental catastrophe unless drastic measures are taken to curb global warming. The impacts of 1.5°C of global warming will be far greater than expected. The report concluded that it is still possible to limit the temperature rise to 1.5°C (the target set by the 2015 Paris Agreement) if the use of fossil fuels is reduced significantly. Human-caused CO_2 emissions would need to be reduced to 45 per cent of their 2010 levels by 2030, and to zero by mid-century.

Predicted environmental and economic impacts include:
- extreme weather events and heat-related morbidity and mortality — extremely hot days, such as those experienced in the northern hemisphere in the summer of 2018, will become more severe and common, increasing heat-related deaths and causing more forest fires
- threatened ecosystems — insects, which are vital for pollination of crops, and plants are almost twice as likely to lose half their habitat at 2°C compared with 1.5°C; corals (p. 12) would be 99 per cent lost at the higher of the two temperatures, but more than 10 per cent have a chance of surviving if the lower target is reached
- coastal flooding — sea-level rise would affect 10 million more people by 2100 at 2°C compared with 1.5°C
- sea-ice-free summers in the Arctic — these would occur once every 100 years at 1.5°C, but every 10 years at 2°C
- reduced marine fisheries catches — these would be 3 million tonnes lower at 2°C, twice the decline at 1.5°C, due to elevated acidity and lower levels of oxygen

Feedback refers to effects that either amplify (positive feedback) or diminish (negative feedback) a change.

Exam tip

Feedback is one of the specialised concepts. You will earn good AO2 credit for integrating these concepts into your responses. Think of other specialised concepts you could mention in a discussion of shifting climate belts.

The **tipping point** is the theoretical point after which the effects of climate change become irreversible.

Knowledge check 45

In what ways will taiga (boreal) forests impact the global climate through radiation balance and carbon cycling?

Other predicted impacts include large-scale singular events (e.g. ice sheet collapse), river flooding, decreased crop yields and increased water restrictions, highlighting how economic and environmental impacts are interrelated.

Strategies to mitigate and adapt to climate change

Mitigation and adaptation are not alternatives; they will have to operate together. Adaptation refers to people and societies who change their lifestyles to cope with a new environment rather than trying to prevent climate change. Examples include:

- using GM technology to develop drought-resistant crops
- managing the coastline sustainably in areas vulnerable to sea level rise, for example by replanting mangroves or managing natural retreat (realignment)
- investing in various forms of water storage to provide safe and clean water supplies to cope with more frequent droughts

Adaptations tend to happen at a local scale, because they are tailored to specific local impacts, and use all levels of technology.

Mitigation refers to a reduction in the output of greenhouse gases and/or increasing the amount of greenhouse gas storage. Examples include:

- setting targets to reduce greenhouse gas emissions, usually by developing a global framework, which is then translated into national strategies
- switching from fossil fuels to recyclable and renewable sources of energy
- capturing carbon emissions and/or storing or burying them, for example in old oil wells

Mitigation needs to operate at a variety of scales — personal, local, national and global.

Summary

- The atmosphere is structured into layers, with most weather processes taking place in the troposphere. Global atmospheric circulation gives rise to the world's high- and low-pressure belts and surface winds. Climate is also influenced by oceanic circulation and altitude.
- The world's main climate types in tropical and temperate regions have distinctive characteristics of temperature, precipitation, wind and pressure. Seasonal variations of climate occur due to the Earth's axial tilt and its orbit around the sun.
- Extreme weather events are becoming more of a feature of the Earth's climate. Populations with greater exposure to climatic variability, greater sensitivity to stress and lower adaptive capacity are more vulnerable to such events.
- Low- and high-pressure systems are associated with different hazards, which impact on the environment and human activity. There is a range of strategies used to manage these.
- Human activities impact on urban climates and air quality, and strategies are needed to reduce these impacts.
- Anthropogenic climate change is causing climate belts to shift. Tipping points occur where the effects of climate change become irreversible. Strategies are required to mitigate and adapt to climate change.

Questions & Answers

Six sample questions are provided, one for each of the optional themes in Section B of the Eduqas and WJEC examinations on *Contemporary themes in geography*. The questions that follow are typical of the style and structure that you can expect to see on your exam papers. They are all extended-response/essay questions, so only extracts of sample student responses are provided, rather than entire answers.

Exam comments

Each question is followed by a list of suggested indicative AO2 content that a good answer should include to provide evidence of the student's ability to apply their knowledge and understanding (icon ⓔ). You should aim to consider additional AO2 content that would be appropriate. You should be able to obtain relevant AO1 content for each question from the content guidance section of this book and from your own research. In order to reach the highest bands of the mark scheme a learner need not cover all the points mentioned in the indicative content, but must meet the requirements of the highest mark band of the generic mark scheme for the respective paper. All student responses are then followed by exam comments. These are preceded by the icon ⓔ and indicate the band that would be awarded for AO1 and AO2, together with any suggested improvements, in some cases using numbered references to indicate the points that specific comments refer to. The AO3 element for all six answers meets the criteria for the highest mark band (band 5 for Eduqas and band 3 for WJEC). The level of student answers given here would achieve a B or A grade overall.

■ Ecosystems

Question 1

To what extent is moisture the greatest influence on the structure and functioning of ecosystems?

ⓔ Answers might include the following AO2 points:

- Whether water is in the form of ice/snow is also a function of temperature, illustrating that ecosystems rely on a variety of interlinked processes to function effectively, making it difficult to isolate the role of water alone.
- Light is critical for photosynthesis, and temperature controls the rate of plant metabolism, which in turn determines the amount of photosynthesis. Photosynthesis is therefore influenced by light and temperature, but cannot occur in the absence of water.

Extract from student answer

Ecosystems contain both biotic (living) and abiotic (non-living) components, which are interconnected and have particular roles. Water is an abiotic and limiting factor for primary productivity, as well as influencing amounts and rates of decomposition and weathering of soils to release nutrients for plant growth. This essay will examine the extent to which water and other factors influence the structure and functioning of ecosystems.

Primary ecological productivity can be defined as the rate at which energy can be converted into organic matter. It can be measured by the amount of new biomass produced each year. Although moisture is a key component in many chemical reactions in plants, ecological productivity also depends on temperature, which speeds up the rate of chemical reactions, light for photosynthesis and nutrient availability for plant growth. Where these factors occur in abundance, levels of primary productivity will be high, but if any one of the four factors is missing, this will limit primary productivity...

...Water is a principal requirement for photosynthesis and the main chemical component of most plant cells. It is therefore critical for both ecosystem structure and functioning. Areas with limited water supply are associated with ecosystems of low productivity such as desert and tundra biomes. In areas of similar temperatures, such as within the tropics, high and constant moisture levels are associated with biomes such as the tropical rainforest, whereas low and/or seasonal moisture availability are associated with savanna grasslands and deserts...

...Nutrients are the chemical elements and compounds required for organisms to grow and function. Nutrient availability is therefore an important limiting factor. However, the fact that nutrients enter the nutrient cycle via precipitation and through the weathering of parent material, a process that is partly dependent on moisture availability, demonstrates that all the limiting factors are in some form of relationship with each other — they are interdependent and it is therefore extremely difficult to isolate one limiting factor alone...

...An influence of growing importance in the structure and functioning of ecosystems is human activity, particularly in terms of the control of limiting factors. On a local level, moisture levels can be controlled through irrigation, nutrient levels through the application of fertilisers and temperature and light levels through artificial heating and lighting in greenhouses. On a regional and global level, these limiting factors cannot be so easily manipulated, although human activity is influencing temperatures through climate change...

...On balance, this essay suggests that, although important, water does not exert the greatest influence on ecosystem structure and functioning. Ecosystems are comprised of biotic components influenced by abiotic ones, one of which is water. Without the sun's input of energy, photosynthesis cannot take place; therefore light may be considered as the most important influence on ecosystem structure and functioning. Also, without the sun's input of energy, neither the atmospheric system nor the hydrological cycle can function.

Therefore, solar energy can be regarded as the most important factor. Moisture, together with temperature and nutrient availability, are all important but cannot be considered to have the greatest influence.

AO1 band 5 (Eduqas), band 3 (WJEC) Although the response above contains only extracts from the student's full answer, the knowledge of ecosystem structure and functioning shown is thorough and conceptual understanding is of a high order. The student has an explicit grasp of ecosystem terms, such as primary productivity and nutrient cycling. Providing some statistical support, such as amounts of primary productivity and precipitation totals, would earn the student additional credit.

AO2 band 5 (Eduqas), band 3 (WJEC) The student demonstrates the ability to express complex arguments effectively. The answer demonstrates the complexity of the relationships between moisture and other limiting factors — not only light, but also temperature and nutrient availability — using the tropical rainforest, tundra and desert biomes by way of illustration. Anthropogenic issues are explored at varying spatial scales (local to global). There is reference to both synoptic links (climate change) and specialised concepts (causality, interdependence).

■ Economic growth and challenge: India

Question 2

Evaluate the view that recent changes in the size and structure of India's economy are primarily due to globalisation.

ⓔ Answers might include the following AO2 points:

- Causal factors are interdependent. Globalisation is linked to government policies (political factor) and has been reinforced by India's colonial heritage (social factor).
- Changes over time. The transition from a highly protectionist and anti-export nation after independence from Britain to an open and globalised economy as a result of the 1991 debt crisis was triggered by political factors, with globalisation assuming greater importance over time.

Extract from student answer

The agricultural sector has undergone several changes, which have contributed to India's thriving economy. Traditional agriculture is in decline as more and more land is being urbanised or bought up by major agricultural TNCs such as Monsanto. Agribusiness has seen a rise, with more farms being run on strictly commercial principles, increasing food production considerably and raising the revenues generated by this sector. Therefore, in the case of the agricultural sector, globalisation has been an important causal factor in its growth, but this was contingent on the Indian government opening the economy to FDI following the 1991 debt crisis. Also, as each state influences the delivery of agri-services, technology and investment, there are spatial variations in the role of globalisation in encouraging the growth in agriculture. Punjab, for example, plays an important role in Indian grain production, with more than 83 per cent of the state under intensive agriculture. Globalisation and political factors are therefore inextricably linked, and the role of environmental factors that promote agriculture in certain parts of India cannot be overlooked. a

However, the agricultural sector's contribution to Indian GDP decreased from 32% in 1990 to 15% in 2017, whilst the contribution of the service sector rose to 59%. This illustrates that it is the tertiary sector that has mainly been responsible for fuelling India's economic growth, with India described as the 'global outsourcing capital'. This term suggests that globalisation has been the critical factor in accounting for economic expansion. Closer analysis reveals a range of interrelated causes. The relative cost advantage of the Indian workforce, which is just as skilled and equipped as that of British and American companies (12% of their 1.1. billion population have a university degree), but much cheaper to employ, has attracted global companies to set up call centres in India. The prevalence of English as a global language has been critical in accounting for the attraction of global companies such as BT to India but could not have happened without the historical influence of India's colonial past. Also, investment has been concentrated in urban hubs such as Bangalore, rather than remote rural areas such as Bihar and Orissa, where educational levels and access to services are lower, which would discourage FDI. In time, as the size of India's consumer base grows, in contrast to the stable or contracting populations of most high-income countries, other parts of India's tertiary sector are set to expand, such as the entertainment industry (Bollywood). Although globalisation will play a part in this, it is India's large and growing population (i.e. demographic factors) that will be the major driver of growth. b

e a In this paragraph the student establishes the link between changes to the primary sector of India's economy and globalisation (AO1). The paragraph is discursive, with the application of specialised concepts of causality, globalisation and interdependence, together with a recognition of variations over time and space. Both the AO1 and AO2 elements are strong (Eduqas Band 5, WJEC Band 3), with specific facts (changes in Indian agriculture over time) integrated with discursive points, including the pivotal role of government decisions to liberalise India's economy and exemplar support to reinforce arguments. Some specific detail about the environmental factors that promote agriculture (locations, climate, soils) would have earned additional credit.

b Here the student provides some specific detail of structural changes to India's economy (AO1) and links these to causal factors, particularly globalisation. The student recognises, however, that globalisation has not occurred in isolation, and provides additional contributory and interrelated factors, again recognising spatial variations in globalisation's role. The final sentences look at part of the tertiary sector — the entertainment industry (AO1) — and attribute its growth more to demographic factors than globalisation (AO2), with synoptic links to population growth rates in other parts of the world.

■Economic growth and challenge: China

Question 3

'China's physical environment provides more opportunities than constraints for economic development.' Discuss.

ⓔ Answers might include the following AO2 points:

- Opportunities and constraints vary spatially — water resources are more abundant to the south.
- Some opportunities may create constraints — the combustion of China's abundant coal reserves, which has have fuelled economic growth, has led to increases in the cost of environmental amenity and repair, placing a strain on China's economy.

> **Extract from student answer**
>
> ...China benefits from large reserves of coal, which have been used to fuel its economic development. However, these are unevenly distributed and there are limited reserves of high-quality coking coal and anthracite, and both these issues act as a constraint. Although the combustion of China's low-grade coal powers 'the workshop of the world', it also increases emissions of CO_2, leading to increased costs of environmental amenity and repair, and threatening the sustainability of China's economic growth...
>
> ...Therefore, China's physical environment presents both opportunities and constraints, and the balance between the two depends on the aspect of the environment under consideration, its location and changes over time. Coal has provided the basis for China's rapid industrialisation, but the constraints associated with fossil fuel use, such as damage to the environment and people's health, are becoming more apparent over time. Conversely, some constraints, such the inaccessible nature of the Qinghai-Tibet Plateau, are now recognised as presenting opportunities for tourism. On balance it can be argued that unless China's physical environment is managed more effectively, for example by using the country's solar and hydropower potential in place of its fossil fuel base, the opportunities will be outweighed by constraints, preventing a sustainable future.

AO1 band 4 (Eduqas), band 2 (WJEC) Although the response above contains only extracts from the student's full answer, the knowledge of China's physical environment shown in the first paragraph lacks specific locational support and examples. It is worth bearing in mind that an annotated sketch map outlining some of the locations of key resources, landscape and concentrations of hazards (typhoons, earthquakes, sandstorms) may save writing time and earn good AO1 and AO2 credit.

AO2 band 4 (Eduqas), low band 3 (WJEC) The student provides accurate application of knowledge and understanding, but only if this is supported by more detailed evidence (see above) can the student achieve the top band. There is reference to specialised concepts of causality (coal fuelling economic growth) and sustainability, and an appreciation of variations over space and time. The argument of the opportunities presented by coal becoming a constraint over time is repeated — the student's time would have been more effectively spent covering other aspects of China's physical environment.

■ Economic growth and challenge: development in an African context

Question 4

'Strategies to manage desertification are not effective.' With reference to two or more sub-Saharan African countries, to what extent do you agree?

ⓔ Answers might include the following AO2 points:

■ The interdependence of strategies to manage both the causes and consequences of desertification, such as forestry management to prevent soil erosion and increase the water-holding capacity of the soil.

■ Different strategies (forestry management and introducing solar and wind power to reduce dependence on wood as an energy source) achieve similar objectives.

Extract from student answer

By 2020, it is estimated that 50 million people in sub-Saharan Africa will be forced to migrate because of desertification, the process by which fertile land becomes desert. Desertification has multiple causes, including climate variability, and a variety of consequences, such as declining agricultural productivity. Strategies to manage both the causes and consequences of desertification are therefore interlinked and come at a variety of scales... a

...The Productive Safety Net Programme aims to prevent the long-term consequences of short-term food deficits, such as the famine resulting from the 2015 drought in Ethiopia. With the support of the UN, Ethiopia's government provides chronically vulnerable households with 6 months of food assistance annually. In addition, 3 months of emergency food aid is provided to mitigate any unpredicted shocks, ensuring people can survive periods of famine. This strategy, not only implemented in Ethiopia, is an example of bottom-up aid, which addresses both the causes and consequences of desertification. Only 15% of Ethiopia's land surface is arable; this leads to overcultivation in order to provide for a population growing at 2.5% a year. This growing population is more vulnerable to future drought as desertification is forcing crop production onto more marginal land and, with 95% of crops rain-fed, precipitation is the most important determinant of crop yields. Therefore, the Productive Safety Net Programme is an effective strategy for its 8 million participants, but the numbers affected by desertification are much greater... b

> ...Another approach to managing desertification is the Great Green Wall. This regional-scale strategy spans the Horn of Africa from Senegal to Djibouti. The programme is funded by the World Bank, aided by the political will of national governments, and involves planting drought-resistant Acacia trees in a band 15 km wide by 8000 km long, to manage the consequences of desertification. Collectively, 15 million hectares of desertified land has been reclaimed by this strategy, which has been effective in managing both the causes and consequences of desertification. Its success is due to the local stabilisation of rainfall patterns (a physical cause) by increasing evapotranspiration, but it also drives the culture of improved management (a human cause). It is a labour-intensive strategy, providing employment (addressing a consequence) and land management training for agricultural-based subsistence farmers, and focusing on long-term sustainability. However, the success will vary according to the political will and effectiveness of the governments responsible, and participation by local people. **c**
>
> In conclusion, both the Productive Safety Net Programme and the Great Green Wall were initially implemented as strategies to manage the consequences of desertification: food insecurity and the loss of fertile land. As they both focus on the long-term, they have also been successful in managing the causes of desertification too. Although not all schemes are effective, the partial successes of the aforementioned schemes indicate that it cannot be said that strategies to manage desertification at all times and in all places are ineffective. **d**

e **a** The introduction defines the term desertification, and establishes a link between the causes and consequences of desertification (AO1). The paragraph is discursive, with the application of specialised concepts of causality and interdependence, together with a recognition of variations by scale. Both the AO1 and AO2 elements are good (Eduqas band 5, WJEC band 3) with a key fact (the number of people affected by desertification) integrated, but reference to the effectiveness of strategies (the key element of the question) is missing.

b This paragraph provides some specific detail of a national strategy (AO1), which addresses both the causes and consequences of desertification and makes synoptic links to climate change (AO2). Specific details are given of the numbers supported by this strategy, provided as a measure of its effectiveness (AO2).

c Variations in the scale of strategies are demonstrated by using the example of the large-scale Great Green Wall (AO1) and how, by addressing the consequences, it is a sustainable scheme that addresses causes too, although with spatial variations in its effectiveness (AO2).

d The student synthesises (AO2) the arguments and facts presented in the main body of the response to draw an overall conclusion.

■ Energy challenges and dilemmas

Question 5

'The most important physical factors determining the supply of energy are climatic.' Discuss.

ⓔ Answers might include the following AO2 points:

■ The main physical factor influencing the supply of energy varies over time. Whereas geological controls have been critical in the past due to the world's dependence on fossil fuels, climatic factors are increasing in importance with growing energy generation from renewable sources.

■ The main cause may vary according to the scale of analysis. At the global scale geological factors may be the most important, but at regional and local levels it may be climate, for example in terms of high precipitation levels generating large amounts of hydropower.

Extract from student answer

...Geological factors are less important than climatic ones when considering the supply of alternative energy sources such as solar, wind, hydropower and biomass. In these cases, climatic factors are more significant. Although these resources constitute a smaller percentage of the global energy mix, they are often important in supplying local-scale energy in some countries and to some communities. Many nations and regions that experience high insolation take advantage of this to generate sustainable solar energy, such as the Ivanpah solar farm, a 370 MW facility on the Nevada/California border...

...Biomass, ultimately dependent on climatic factors such as the amount of precipitation, insolation rates and temperature, is an important source of energy in many less-developed countries that cannot access global energy markets or cannot invest in a national grid to support their populations. In these cases, local-scale energy generation from biomass is used for domestic and local industrial processes. For example, in Botswana over 90 per cent of energy generation is from informal, local-scale burning of biomass.

Although geological factors were the most important physical factor determining global energy supply in the past, with fossil fuels accounting for over 70 per cent of the global energy mix, in many areas climatic factors are of growing and, in some locations, unparalleled significance. As global reserves of oil, gas and coal are depleted, perhaps by as early as 2100, energy generated from climatic sources is likely to become much more important if sustainability and energy security are to be achieved. Furthermore, if long-term climate change is to be mitigated, a move to energy sources generated from the sun, wind and water that do not release CO_2 is necessary. For these reasons, climatic factors are an increasingly important determinant of energy supplies, and will become more important in the future. However, climatic factors are not the most important physical factor determining the supply of energy in all places and at all times.

AO1 band 5 (Eduqas), band 3 (WJEC) Although the response above contains only extracts from the student's full answer, the student has demonstrated wide-ranging, thorough and accurate knowledge. For example, they provide a link between biomass energy sources and climatic factors, with appropriate and accurate exemplification of locations where biomass energy is important and some indication of how much energy is derived from this source. This high order of conceptual understanding and effective use of geographical terminology was demonstrated throughout the student's answer. Appropriate, accurate and well-developed exemplification always earns good credit.

AO2 band 5 (Eduqas), band 3 (WJEC) This extract shows the application and integration of the specialised concepts of causality (causes of biomass energy sources), inequality (unequal access to global energy markets and energy grids), interdependence (between plant growth and climatic factors), mitigation (of climate change), resilience (energy security) and sustainability (of energy sources), together with a recognition of variations over different time and spatial scales. Synoptic links are made with the carbon cycle.

■Weather and climate

Question 6

'Human activities in urban areas affect temperature more than any other climate variable.' Discuss.

Answers might include the following AO2 points:

■ The term 'urban heat island' indicates that temperature is particularly affected in urban environments.
■ Climatic variables are interdependent: temperature influences relative humidity, possibly leading to convectional rainfall and increased cloud cover.

Extract from student answer

Human activities have a profound influence on urban microclimates. Through altering land uses and concentrating human activities in a relatively small area, all climatic variables, including temperature, wind, precipitation and humidity, are modified. The term 'urban heat island' may indicate that temperature is the climatic variable that is most affected. However, although temperature is always affected by human activities in urban environments, it cannot be viewed in isolation because it acts as a catalyst for changes to other climatic variables. In most mid-latitude European and North American cities, average minimum winter temperatures are 1–2°C higher than in rural environments. This is because urban environments are made up of heat-retaining materials with better radiation-absorbing properties and lower albedo rates. Heat is released from buildings, vehicles and industry. Also, due to the absence of vegetation, less solar energy is used up in evapotranspiration, so more is available to heat the surface. Annual mean wind speeds are around 20 per cent lower and the frequency of extreme gusts around 15 per cent lower than in rural environments.

The incidence of calms is around 20 per cent higher. These statistics can be explained by the frictional drag of the tall buildings in an urban landscape, which reduces wind speeds. Calm conditions also contribute to a higher incidence of fogs, illustrating the interdependence of climatic variables. However, in some cities tall buildings funnel the wind, increasing wind speeds. Total precipitation is 5–30 per cent more; the number of rain days is 10 per cent more and snow days are 14 per cent less than in rural environments. This is because the greater amounts of dust in urban areas increase the concentration of hygroscopic particles that provide condensation nuclei, leading to more rain formation and higher temperatures, resulting in increased convectional rainfall but less snow, again illustrating the interdependence of climatic variables. Relative humidity is 20 per cent less in winter and 8–10 per cent less in summer than in rural environments because the lack of water bodies, less vegetation and higher temperatures (at higher temperatures the atmosphere has a greater capacity to hold water in its vaporous state than at lower temperatures) in urban environments result in lower relative humidity.

It can be argued that although human activities influence the climatic variable of temperature, this in turn will directly or indirectly influence the other climatic variables of humidity, precipitation and wind, illustrating the interdependence of climatic variables. However, given the variety of city sizes, land uses, human activities, regional climates and climate change, the climatic variable most affected will vary from city to city and over time.

A01 low band 5 (Eduqas), band 3 (WJEC) Although the response above contains only extracts from the student's full answer, the description and explanation of changes to climate variables in cities due to human activities is very clear and thorough. Unfortunately, however, in not referring to actual cities, it lacks specific locational support. It is worth bearing in mind that examiners are aware of the time constraints that students are under and would give good credit for the knowledge and understanding shown, but some specific geographical detail would lift the response to the top of the top band.

A02 low band 4 (Eduqas), low band 3 (WJEC) The answer provides accurate application of knowledge and understanding, but this is concentrated in the introduction and conclusion. The specialist concepts of causality (heat-retaining properties of buildings influencing temperatures) and interdependence (higher temperatures resulting in more convectional rainfall) are occasionally made explicit, and some synoptic elements are evident (convectional rainfall and climate change from the water and carbon cycles). An appreciation of variations over space and time is evident in the conclusion. However, the knowledge and understanding, particularly in the second paragraph, needs to be applied more to the question, and evidence in support needs greater range and depth to achieve the top band.

Knowledge check answers

1 Links might include: carbon and water — the net accumulation of carbon over time in undisturbed peatland provides an important carbon store; changing places — the traditional foods and clothing provided by locally grown crops and from animals make a place distinctive and give it place identity; weather and climate — mangrove forests and coral reefs protect shorelines from the impact of tropical low-pressure systems.

2 Coral reefs would be one example through destructive fishing practices, the over-exploitation of fish, marine pollution and runoff, rising temperatures and ocean acidification.

3 In the temperate grassland biome, because the low temperatures in winter cause the grasses to die back to their roots and photosynthesis (a process that sequesters carbon) ceases. Carbon exchanges are rapid throughout the year in tropical rainforests due to the availability of light, high temperatures and heavy rainfall.

4 See figures 6 and 7 (p. 10). The cold winters and short growing season result in a small biomass store. Large numbers of bacteria return nutrients from the litter to the soil, which is the largest store. The relatively dry climate prevents the loss of nutrients through leaching. The nutrient-rich soil is very fertile, supporting arable farming, for example the Canadian Prairies.

5 Due to the distance from the sun and the curvature of the Earth, the amount of insolation received at the Earth's surface decreases in quantity towards the poles. The northern hemisphere receives its maximum amount of insolation between March and September, when it is tilted towards the sun, but when the northern hemisphere is tilted away from the sun during the winter months, the polar region receives little to no insolation.

6 The tundra plants are dark and hairy. The darkness of their flesh absorbs solar heat, and the hairs help to trap the heat and keep it close to the surface of the plant. Some plants even have dish-like flowers that track the sun. Animal adaptations include: short and stocky arms and legs; thick, insulating cover of feathers or fur, which changes colour (brown in summer, white in winter); thick fat layer gained quickly during spring in order to have continual energy and warmth during winter months; chemical adaptations in some animals to prevent their bodily fluids from freezing.

7 The number of children (aged under 15) and old people (aged 65 and over) as a ratio to the number of adults (aged between 15 and 64). It indicates the number of people who the economically active population supports.

8 India has a low and declining female:male gender ratio. The ratio among Hindus is the lowest. India's patriarchal structure skews the sex ratio in favour of males. A son is perceived as an 'asset', because he traditionally supports the family in an agrarian-based society. A daughter is perceived as a 'liability', because traditionally her family raises a dowry for her marriage. Practices of female foeticide (despite a ban on the abortion of female foetuses) and infanticide, together with female discrimination, and high maternal and female mortality, persist. Problems associated with this gender imbalance include rising crime rates and the development and growth of 'bachelor' villages.

9 In addition to environmental concerns (coal is one of the dirtiest hydrocarbon fuels), coal cannot meet all of India's energy needs. The transportation industry needs oil, and much of India's coal is not of the type needed in steel and other industries.

10 The Himalayas are a good example of a continental-to-continental collision (or destructive) margin. The Indo-Australian plate is moving northwards at a rate of 5–6 cm/year, so colliding with the Eurasian plate.

11 Other factors include: the emergence and investment policies of MNCs; the growth of an urban, educated, middle-class population whose members have become consumers themselves, and who provide a large market for new consumer goods.

12 Rapid economic growth has been due to the expansion of the service sector rather than the growth of manufacturing. Business links to North America and Europe dominate rather than those to Japan and other Asian nations.

13 The demographic dividend is the accelerated economic growth that results from a decline in a country's mortality and fertility, and the subsequent change in the age structure of the population to provide a large economically active sector.

14 Wipro or Infosys (both based in Bangalore (Bengaluru))

15 Desalination is expensive, energy intensive and is a location-specific solution because plants require a coastal location.

16 The South–North Water Diversion Project (p. 44).

17 In addition to environmental concerns (coal is one of the dirtiest hydrocarbon fuels), coal cannot meet all of China's energy needs. The transportation industry needs oil, and much of China's coal is of low quality.

18 Earthquakes are generated by the collision of the Indian and Eurasian tectonic plates. As the Indian plate moves northwards into the Eurasian plate, it forces the Tibetan Plateau eastward into China.

19 The original four special economic zones were sited in coastal areas of Guangdong and Fujian that had a long history of contact with the outside world through outmigration, and at the same time were near Hong Kong, Macao and Taiwan. The choice of Shenzhen was especially strategic because it is situated near Hong Kong, the key area from which to learn capitalist modes of economic growth.

20 Examples include DJI Innovations, Haier, Lenovo and Datang Telecom.

21 Other factors include: the emergence and investment policies of MNCs; the growth of an urban, educated, middle-class population who provide a large market for new consumer goods; improved education; and a rapidly improving transport infrastructure.

22 Costs include: skilled jobs being filled by imported Chinese labour; Chinese aid to African countries being tied aid; factories being Chinese-owned and preventing indigenous firms from developing; and preventing the development of a manufacturing base in Africa (the export of raw materials from African countries still dominates trading patterns). Benefits include aid, employment opportunities, increased capital, technology and expertise.

23 Youthful out-migration has left the countryside with an ageing workforce. This threatens future agricultural production and China's food security.

24 Nanhui is one of China's 285 purpose-built ecocities.

25 Low-income economies are defined as those with a GNI per capita of US$995 or less in 2017. Of the 34 countries listed in this category in 2019, 27 are in sub-Saharan Africa.

26 The difference in rankings reflects that some countries are more 'developed' in some aspects of development than in others.

27 The East African Rift Valley — a 2000-mile-long constructive plate margin — runs through Kenya.

28 MDGs — eight international development goals for the year 2015 that had been established following the Millennium Summit of the United Nations in 2000.
SDGs — 17 global goals set by the United Nations General Assembly in 2015 to address poverty, inequality, climate, environmental degradation, prosperity, peace and justice.

29 Bare, dry ground provides a lighter surface than the darker vegetation cover it replaces. Albedo therefore increases and the ground is not heated as intensely. A cooler surface results in less convectional rainfall and reduced precipitation totals.

30 Infiltration-excess overland flow occurs when precipitation intensity exceeds the infiltration capacity of the soil, whereas saturation-excess overland flow occurs when the soil becomes saturated, and any additional precipitation causes runoff.

31 LNG is liquified natural gas. With North Sea gas fields in decline, an increasing supply of natural gas in the UK comes in the form of LNG. Chilled to -160° and shipped across oceans, it is argued that carbon emissions associated with LPG are higher than for shale gas.

32 The main players in the global energy market are MNCs, OPEC, national governments, energy companies and consumer and pressure groups.

33 Carbon neutral means having a net zero carbon footprint. It is achieved by balancing the amount of carbon released with the amount sequestered.

34 In Iceland electricity is generated entirely from renewables (75 per cent from hydropower and 25 per cent from geothermal), but with the balance changing in favour of geothermal, reflecting the dynamic nature of energy supply.

35 A range of countries have developed or are embarking on nuclear power programmes, ranging from sophisticated economies (e.g. France) to middle-income (e.g. India) and developing nations (e.g. Nigeria). Iran's nuclear power programme has caused political concerns because of the potential it creates for developing nuclear weapons.

36 More outgoing terrestrial radiation will be absorbed and re-radiated back to the Earth's surface, causing further warming.

37 Warm currents, because air that blows over a powerful warm current, such as the North Atlantic drift, is warmed up and raises temperatures when blown onshore.

38 On western land margins, cold ocean currents deprive inblowing marine winds of their moisture before they reach land, thus contributing to exceptionally dry conditions and leading to fog conditions offshore, for example off northern Chile and Namibia.

39 Winnipeg, which has a continental location. Land surfaces heat up and cool down rapidly in comparison with maritime locations. Therefore, Winnipeg experiences a much greater range in temperatures than Vancouver, which is located on the coast.

40 Convection rainfall is rain that occurs due to the heating of the ground, causing warm air to rise, cool and condense. The two other types are frontal and relief (orographic) rainfall.

41 Due to differential heating between land and sea

42 One reason would be because of the decrease in snow and ice cover, which lowers albedo. More solar energy is absorbed by the Earth's surface, leading to further heating, i.e. the rate of heating is increased. A second reason is that, as temperatures increase, permafrost will thaw, releasing large quantities of CO_2 and methane (a potent GHG), leading to further warming.

43 Acid rain has regional impacts well beyond the primary source of pollution.

44 Because car pollution is trapped beneath temperature inversions formed under winter anticyclones.

45 The replacement of tundra by more boreal forests will lead to higher levels of primary productivity and increased CO_2 uptake, but CO_2 emissions from thawing permafrost will likely counterbalance this positive impact. Boreal forests create darker surfaces than the tundra they replace, leading to reduced albedo and greater absorption of solar insolation.

Index

Note: **bold** page numbers indicate key term definitions.

Index